PROGRESS WITHOUT PEOPLE

DAVID F. NOBLE

PROGRESS WITHOUT PEOPLE

NEW TECHNOLOGY,
UNEMPLOYMENT,
AND THE
MESSAGE OF RESISTANCE

between the lines

© David F. Noble, 1995

Published by:
Between The Lines
720 Bathurst Street, #404
Toronto, Ontario
Canada   M5S 2R4

Cover Design: Counterpunch
Interior: Steve Izma
Printed in Canada
Second printing June 1997

Between The Lines gratefully acknowledges financial assistance from
Canada Council
Canadian Heritage Ministry
Ontario Arts Council

All rights reserved. No part of this publication may be reproduced, stored in a retrieval system, or transmitted in any form by any means, electronic, mechanical, photocopying, recording, or otherwise, except as may be expressly permitted in writing by the publisher or CANCOPY (photocopying only), 6 Adelaide Street East, Suite 900, Toronto, Ontario, M5C 1H6.

Canadian Cataloguing in Publication Data

Noble, David F.
   Progress without people: new technology, unemployment, and the message of resistance

ISBN 1-896357-01-6 (bound)   ISBN 1-896357-00-8

1. Labor supply — Effect of technological innovations on. 2. Technological innovations — Social aspects. 3. Luddites. I. Title.

HD6331.N62 1995   303.48'3   C95-930762-1

For Doug

Contents

Preface ix
Introduction xi

Part One: Another Look at Progress 1
1 In Defence of Luddism 3
2 The Machinery Question Revisited 24
3 Present-Tense Technology 40

Part Two: Automation Madness: Or, the Unautomatic History of Automation 69
4 Automatic Technological Progress 71
5 A Second Look at Social Progress 101
6 The Hearings on Industrial Policy: A Statement 115
7 The Religion of Technology: The Myth of a Masculine Millennium 127

Appendices 143
I Nineteenth-Century Consultant to Industry Saw Automation as Weaponry 145
II Karl Marx against the Luddites 146
III A Technology Bill of Rights from the International Association of Machinists 147

IV	"Starvin' in Paradise" with the New Technology	150
V	Lord Byron Speaks against a Bill to Introduce the Death Penalty for Machine-Breaking	154
VI	An Exchange between Norbert Wiener, Father of Cybernetics, and Walter Reuther, UAW President	161

A Note on the Author 165

PREFACE

Most of the chapters in this book were written a decade ago at a time of unprecedented corporate restructuring and, hence, vulnerability—a rare but now lost opportunity for effective opposition. They were written in the vain hope that labour organizations might creatively seize the moment and turn the tide to their advantage before capital was able to consolidate its gains.

Part One, the chapters of which first appeared as articles in the U.S. journal *democracy* in 1983, is an attempt to account for the lack of full-scale labour resistance to the corporate technological offensive by means of a historical analysis of our inherited ideology of technological progress. Part Two is drawn from talks delivered to labour audiences in Canada, Europe, Japan, and the United States between 1982 and 1986. At the time, the message of resistance was being effectively marginalized not only by aggressive management propaganda campaigns (and extortion and repression) but also by labour's own desperately hopeful allegiance to the agenda of competitiveness, amidst absurdly optimistic academic appraisals of the alleged promise of computer-based technologies. Added to these talks is a more recently written chapter on the "religion of technology," tracing the religious roots of our irrational faith in technological salvation.

Needless to say, the last decade has proved a catastrophe for labour, and a more sober assessment of the damage and the prospects has begun belatedly to emerge from the debris. The moment, admittedly, is late, but late is still better than never. Perhaps now, in the wake of this unending tragedy, the message of resistance might once again be heard, despite the constant corporate propaganda that continues unabated and unabashed.

D.N.
Toronto
Autumn 1994

INTRODUCTION

This is not yet another forecast of the social impact of information technologies. It is rather a call to action on the basis of what we already know, and an attempt to explain our inaction to date, despite what we know. For there is no need for futuristic speculation about the information highway or federally funded research on the virtual workplace to see what has been happening to our lives and livelihood in the so-called information age. For this electronic epoch is now already half a century old (the terms automation and cybernetics were coined in 1947), and the returns are in. The catastrophe of the second industrial revolution already rivals that of the first, only without the resistance.

The information highway is barely under construction, the virtual workplace still largely experimental, but their consequences are readily predictable in the light of recent history. In the wake of five decades of information revolution, people are now working longer hours, under worsening conditions, with greater anxiety and stress, less skills, less security, less power, less benefits, and less pay. Information technology has clearly been developed and used during these years to deskill, discipline, and displace human labour in a global speed-up of unprecedented proportions. Those still working are the lucky ones. For the technology has been designed and deployed as well to tighten the corporate stranglehold on the world's resources, with obvious and intended results: increasing dislocation and marginalization of a large proportion of the world's population — within as well as without the industrial countries; growing structural (that is, permanent) unemployment and the attendant emergence of a nomadic army of temporary and part-time workers (the human complement of flexible production); a swelling of the ranks of the perpetually impoverished; and a dramatic widening of the gap between rich and poor to nineteenth-century dimensions.

At the same time, for this is simply the reverse side of the same coin, there has been a greater concentration of wealth, and a greater concentration of power in the hands of the world's economic, political, media, military, and intelligence elites. In their hands — and it is now more than ever in their hands — the latest incarnations of information technology will only compound the crime. Given the lack of resistance, much less coherent and cohesive opposition, there is simply no other possibility. However empowering the new technology might sometimes seem, the appearance is deceiving, because the gains are overwhelmingly overshadowed, and more than nullified, by the losses. In short, as the computer screens brighten with promise for the few, the light at the end of the tunnel grows dimmer for the many.

In Canada these consequences are amply evident. Unemployment has risen each decade of the information age, with the increasing deployment of "labour-saving" technology. In the 1940s, at the dawn of this new age, average official unemployment stood at 2.7 per cent. It rose to 4.2 per cent in the 1950s, 5.1 per cent in the 1960s, 6.7 per cent in the 1970s, and 9.3 per cent in the 1980s. Thus far in the 1990s official unemployment has averaged about 11 per cent, and the new wave of information technology is only just cresting. The official unemployment rate is, of course, an understatement, since those who have given up looking for a job are not included in the count. When they are counted, the rate nearly doubles. A fifth of the employed, moreover, are part-time or temporary workers, whose jobs offer little or no benefits beyond a barely subsistence wage. This escalating unemployment has little to do with business-cycle recessions and recoveries; it is structural, a persistent trend. Recoveries are now "jobless recoveries" — in other words, recoveries for the few. Output and profits rise without the jobs that used to go with them.

Where have all the jobs gone? Ask the printers, postal workers, bank tellers, telephone operators, office workers, grocery clerks, airline reservation agents, warehouse workers, autoworkers, steelworkers, dockworkers—if you can find them. Computer-aided manufacturing, robotics, computer inventories, automated switchboards and tellers, telecommunications technologies—all have been used to displace and replace people, to enable employers to reduce labour costs, contract-out, relocate operations. From the factory to the farm, from the oil refinery to the office, no workplace has been immune to this assault.

For decades much of this job loss remained hidden—except to those displaced—behind the compensatory growth of government employment, but now governments too have avidly adopted the same technologically based labour cost-cutting solutions to their deficit dilemmas, with obvious results: the unemployment disaster has nowhere left to hide. In 1993 an economist for the Canadian Manufacturers' Association estimated that in the previous four years two hundred thousand manufacturing jobs had been eliminated through the use of new technology. That was only in manufacturing, and just the beginning. The latest wave of information technology makes past developments seem quaint in comparison. And still there is no outcry, no plea for protection, no evidence of resistance.

The first industrial revolution was quite different. Untold numbers of people were similarly undone in that calamity, but there was also intense resistance, resistance that persisted and ultimately gave rise to the modern labour movement and its corollary, progressive social legislation. Today the results of that struggle are steadily being eroded before our eyes, as trade unions lose their power and social programs designed to protect people from the ruthlessness of the market are daily dismantled. In short, the second industrial revolution, grounded upon information technologies, is being used to undo the hard-won gains of the first, with nothing in sight to replace them. Why the lack of resistance this time?

Why such deference to the market and reverence for technology, even though we should and do know better? This book suggests that it is more than fear that accounts for such collective paralysis—although that is surely part of it. We are paralysed also by our inherited ideas, some of them invented around the time of the first industrial revolution—a time of widespread opposition to unregulated technological change—precisely to induce such paralysis in the face of technological assault. One was the idea of inevitable and inevitably beneficent technological progress; another was the idea that competitiveness, based upon such technological advance, was the surest guarantee of prosperity. Such notions, though wearing thin, continue to confound the opposition.

At the dawn of industrial capitalism, the English political economist David Ricardo at least acknowledged the misery that machinery was wreaking upon his less fortunate countrymen. As capitalists prospered from their introduction of machinery, working-class families confronted the loss of their livelihoods, dislocation from their communities, poverty, and despair. In the midst of the Luddite uprising recounted in this book, Ricardo sympathized with the outrage and resistance of the workers, including their machine-breaking, on moral and political grounds. As an economist, however, he argued that such opposition to machinery was misguided; while no doubt effective in the short run as a delaying tactic, it would ultimately only further disadvantage workers in the long run by destroying their livelihoods altogether. Certainly these people would suffer grievously in the wake of industrialization, Ricardo noted, but they would suffer even more if they resisted. For despite the sacrifice entailed, mechanization was nevertheless the key to industrial competitiveness and, hence, to the wealth of nations. Workers who yielded to unrestricted technological change risked losing some of their jobs to machinery, but those who refused to yield ran the risk of losing all their jobs to more advanced competitors. However harsh the immediate consequences, then, Ricardo maintained, submission to the *laissez innover* logic of the *laissez faire* market was the only safe bet in a developing industrial world.

Today Ricardo's sombre message echoes anew as simply a given. Everyone assumes, without debate, that resistance to technological change is a sure recipe for competitive doom. In stark contrast to Ricardo's day, there is no open resistance in the age of automation, nor even sympathy for resistance. Yet, ironically, Ricardo's logic has today lost whatever meaning it might once have had. For while it remains true that unregulated technological change continues to cause untold misery for working people in the short run, there is now considerable doubt as to whether workers have anything to gain from the competitiveness of their employers' firms in the long run either.

The reason for this is all too obvious: if new technology has made firms more competitive, it has also made them more mobile and global, more able to play any one country's workforce off against the others in search of cheap, compliant labour. The competitiveness of a footloose firm, a corporation without a country, is no longer tied to the wealth and the jobs of the nation that first fuelled its prosperity. Thus, first the company retools, then it relocates. Hence, today, Ricardo's argument about the long-term benefits of increased competitiveness, despite the short-term tragedies, rings hollow, as short-term losses to technology are invariably followed by long-term losses to increased competitiveness.

Moreover, and unforeseen in Ricardo's day, the realities of multinational enterprise belie the supposed logic of the competitive market. Globe-straddling firms routinely collude to divide up the planet, collaborating with their alleged competitors in joint research, development, production, and distribution arrangements, as well as through interlocking investments and directorates. At the same time, the mammoth scale of transnational enterprise means that an ever-increasing proportion of the world's trade and "competition" takes place within rather than between companies. Here the pursuit of competitiveness and markets is an internally managed affair, rather than a struggle for national economic survival.

In short, Ricardo's faith in the competitiveness of firms as a guarantor of the wealth of nations has become an anachronism. This sobering fact is clearly reflected in the recent proliferation of continental and hemispheric free-trade agreements that open national borders to capital, but not to labour, and thereby institutionalize the new era of transnational corporate hegemony. While multinational firms continue to expect nation-states to subsidize their operations and fight their wars, they have no allegiance or responsibility to the people of any nation. There is thus no reason for workers to place their faith any longer in the promise of competitiveness; the only ones really competing these days, it often seems, are the workers themselves — against each other.

Nevertheless, the chorus about the virtues and imperatives of new technology and competitiveness resounds throughout the land, and the strident voices of authority now include not only those of capitalists and their apologists, but also those of politicians of every stripe, techno-zealots, the ubiquitous media, academics, and even most trade union officials. Propaganda for domestic consumption, the incessant incantation, keeps workers in their place, off balance, desperately striving for a stay of execution, and drives home the major message that they have no choice. They are encouraged to deny what they know from experience, to distrust their own minds and senses. Sober debate is ruled out of order, or is drowned out in the hypnotic hum of a mindless multinational mantra. Resistance, or even talk of resistance, is decried as irrelevant, futile, old-fashioned. Pleas for some rational reflection in this time of crisis are dismissed out of hand, as irrational.

Meanwhile, the rude realities of the workplace daily contradict the chorus. More and more workers have already seen what happens when they patiently allow the new equipment into the shop, responsibly sacrificing jobs today for competitive survival tomorrow: the company thrives and then leaves anyway, or contracts out their work to cheaper labour halfway around the globe. For these workers, at least, the game is now up. The enemy is clear, the fairy tales about the information age and competitive prosperity forsworn, the rage and resentment out in the open. They have learned the hard way that, contrary to what they hear from all the self-anointed, self-serving authorities, resistance is not misguided but more essential than ever, for today it is a struggle not just for the short term but for the long term as well, not merely a response to the immediate threat of job loss but a direct challenge to the multinational marauders. Resistance can hamper a corporation's mobility, restrict its reach, weaken its hold on our lives and futures. But such reawakened wisdom and renewed resolve have been perilously long in coming.

A recent "Futurescape" advertising supplement in *The Globe and Mail* by Rogers, Cantel, and Bell ominously warned that the information highway "raises the ante in competition. If we don't act, Canada and Canadian companies will be left behind. . . . The information highway is not a luxury technology for the rich. It is the way of the future. And those who do not get on the highway will not have any way of reaching their ultimate destination." What precisely is that ultimate destination, the reaching of which now depends upon taxpayer support for this new corporate infrastructure? The propaganda doesn't say, but most people instinctively seem to know anyway. According to a 1993 Gallup poll, 41 per cent of Canadians currently employed believe they will lose their jobs. They are probably right, because for a growing number of people the ultimate guaranteed technologically delivered destination is the dole, whatever is left of it.

At this very moment somewhere today, as this is being written (or as you are reading it), a truck bearing new technology is backing up to a loading dock. The workers gathered to unload and install the new machinery no doubt realize that the new equipment will most likely cost some of them their jobs and give management more control over production and greater power over their working lives. What can they do? What action might they take to protect themselves? And what might the rest of us be doing, to help extend their range of options, secure support for their efforts, and ensure that, whatever their course of action, they need never act alone?

Without answers to these simple yet urgent questions, all discussion about new technology will remain merely academic, and progress without people will proceed apace, carrying us all to our ultimate destination.

ANOTHER LOOK AT PROGRESS

PART 1

CHAPTER 1

IN DEFENCE OF LUDDISM

There is a war on, but only one side is armed: this is the essence of the technology question today. On the one side is private capital, scientized and subsidized, mobile and global, and now heavily armed with military-spawned command, control, and communication technologies. Empowered by the second industrial revolution, capital is moving decisively now to enlarge and consolidate the social dominance it secured in the first.

In the face of a steadily declining rate of profit, escalating conflict, and intensifying competition, those who already hold the world hostage to their narrow interests are undertaking once again to restructure the international economy and the patterns of production to their advantage. Thus, with the new technology as a weapon, they steadily advance upon all remaining vestiges of worker autonomy, skill, organization, and power in the quest for more potent vehicles of investment and exploitation. And, with the new technology as their symbol, they launch a multimedia cultural offensive designed to rekindle confidence in "progress." As their extortionist tactics daily diminish the wealth of nations, they announce anew the optimistic promises of technological deliverance and salvation through science.

On the other side, those under assault hastily abandon the field for lack of an agenda, an arsenal, or an army. Their own comprehension and critical abilities confounded by the cultural barrage, they take refuge in alternating strategies of appeasement and accommodation, denial and delusion, and reel in desperate disarray before this seemingly inexorable onslaught—which is known in polite circles as "technological change."

What is it that accounts for this apparent helplessness on the part of those whose very survival, it would seem, depends upon resisting this systematic degradation of humanity into mere disposable factors of production and accumulation? To be sure, there is a serious imbalance of power between the opposing forces, and perhaps an immobilizing fear on the weaker side in the face of so awesome an assault. But history is replete with examples of how such weaker forces have valiantly defied, and even triumphed over, the stronger. Why then this striking lack of resolve against the new technological offensive?

In search of an explanation for this apparent paralysis, and a cure for it, this book explores beyond the constraints of the current crisis to focus upon older and more fundamental handicaps. Rather than examining the well-known enemies without—the tactics and threats of multinational corporations that are daily reported in the press and chronicled by a spectrum of specialists—this analysis examines the enemies within—the opposition's own established patterns of power and inherited habits of thought that now render it so supine and susceptible. These internal foes, at once political and ideological, can only successfully be overcome by means of direct and frank confrontation, which is the task begun here.

In outline, the opposition suffers from a fatalistic and futuristic confusion about the nature of technological development, and this intellectual problem is rooted in, and reinforced by, the political and ideological subordination of people at the point of production, the locus of technological development. This twofold subordination of workers, not alone by capital but also by the friends of labour (union officials, left politicians, and intellectuals), has hardly been accidental. Rather, it has served the interests of those who wield control over labour's resources and ideas. For the political subordination of workers has disqualified them from acting as subjects on their own behalf, through their own devices and organizations, and thus has minimized their challenge to the labour leadership. And the ideological subordination of the workers has invalidated their perceptions, knowledge, and insights about what is to be done, and has rendered them dependent upon others for guidance, deferential to the abstract and often ignorant formulations of their absentee agents.

Such subordination has handicapped the opposition to the current technological assault in several ways. First, and perhaps most obvious, it has eliminated from the battle those actually on the battlefield of technological innovation, those best situated to comprehend what is happening and to fight effectively. Second, in denying the possibility of people at the point of production participating on their own behalf in the struggle, the opposition has lost as well its understanding of what is actually happening—an appreciation that arises only from daily confrontation, extended experience, and intimate shop-floor knowledge. Finally, the political and ideological subordination of people at the point of production has entailed a removal of the technology question from its actual site and social context, with serious consequences.

While this removal of the technology question has perhaps strengthened the position of the friends of labour, vis-à-vis the workers themselves, it has weakened them vis-à-vis capital. For without any power rooted in the self-activity of the workers at the point of production, the friends of labour have become more susceptible to the power of others. Without a firm grasp of reality based upon experience, they have become abstract in their thinking, and more vulnerable to the ideas of others. (It must be emphasized that this is not a matter of individual integrity or weakness, but rather a powerful cultural phenomenon that has influenced everyone.)

The impotence and ignorance resulting from the double disqualification of people at the point of production, moreover, have manifested themselves in profound intellectual confusion about the nature and promise of technological development itself. Abstracted from the point of production, and therefore from the possibility of a genuinely independent point of view, the opposition's own notion of technological development has come to resemble and ratify the hegemonic capitalist ideology of technological necessity and progress. For it too has become a mere ideological device, an enchantment, and an opiate. The idea of technology has lost its essential concreteness, and thus all reference to particulars of place and purpose, tactics and terrain.

Without moorings in space, the disembodied idea has wandered adrift in time as well. Technological development has come to be viewed as an autonomous thing, beyond politics and society, with a destiny of its own which must become our destiny too. From the perspective of here and now, technological development has become simply the blind weight of the past on the one hand and the perpetual promise of the future on the other. Technological determinism—the domination of the present by the past—and technological progress—the domination of the present by the future—have combined in our minds to annihilate the technological present. The loss of the concrete, the inevitable consequence of the subordination of people at the point of production, has thus resulted also in the loss of the present as the realm for assessments, decisions, and actions. This intellectual blind-spot, the inability even to comprehend technology in the present tense, much less act upon it, has inhibited the opposition and lent legitimacy to its inaction.

This chapter examines the origin of this paralysis and the ideas that sanction it, looking in turn at the first and second industrial revolutions. The aim is to regain the concrete by affirming the perceptions of those at the point of production, thereby to reclaim the present as a locus of action—while there is still time to act. For people at the point of production were the first to comprehend the full significance of the first industrial revolution and to respond accordingly. They have also been the first to see the second industrial revolution for the devastating assault that it is—not because of their superior sophistication at dialectics but because of what it is already doing to their lives—and to respond accordingly.

The purpose here is to acknowledge, endorse, and encourage their response to technology in the present tense, not in order to abandon the future but to make it possible. In politics it is always essential to construct a compelling vision of the future and to work toward it, and this is especially true with regard to technology. But it is equally essential to be able to act effectively in the present, to defend existing forces against assault and try to extend their reach. In the absence of a strategy for the present, these forces will be destroyed, and without them all talk about the future becomes merely academic.

No one alive today remembers first-hand the trauma that we call the first industrial revolution, which is why people are now able so casually to contemplate (and misunderstand) the second. What little we actually know about those earlier times—perhaps the only adequate antecedent to our own—has filtered down to us through distorting lenses devised to minimize this calamity and justify the human suffering it caused in the name of progress. The inherited accounts of this period were formulated by and large in response to the dramatic actions of those who fought for their survival against this progress. They constituted a post hoc effort to deny the legitimacy and rationality of such opposition in order to guarantee the triumph of capitalism.

The Luddites were not themselves confused by this ideological invention. They did not believe in technological progress, nor could they have; the alien idea was invented after them, to try to prevent their recurrence. In light of this invention, the Luddites were cast as irrational, provincial, futile, and primitive. In reality, the Luddites were perhaps the last people in the West to perceive technology in the present tense and to act upon that perception. They smashed machines.

The effort to reconstruct this earlier period of the first industrial revolution might help us to deconstruct our inherited perceptions of technology, because those perceptions date in large part from this historical watershed. Fortunately, during the last several decades and in an effort to understand the opposition to progress in its own terms, social historians have made great strides toward just such a reconstruction. In particular, they have sought to redeem those who have come to seem so irrational and wrongheaded, and they have discovered that the resistance was in fact rational, widely supported, and indeed successful—both in buying time for reflection and strategizing (something today's labour movement would surely welcome) and in awakening a far-reaching political consciousness among workers.

According to these revisionist interpretations, the Luddites who resisted the introduction of new technologies were not against technology per se but rather against the social changes that the new technology reflected and reinforced. Thus, the workers of Nottingham, Yorkshire, and Lancashire were not opposed to hosiery and lace frames, the gig mill and shearing frames, larger spinning jennies, or even power looms. Rather, in a postwar period of economic crisis, depression, and unemployment much like our own, they were struggling against the efforts of capital, using technology as a vehicle, to restructure social relations and the patterns of production at their expense. During the first three decades of the nineteenth century, the workers in manufacturing trades united in opposition to unemployment, the lowering of wages, changes in the system of wage payments, the elimination of skilled work, the lowering of the quality of products, and the factory system itself, which entailed an intensification of work discipline and a loss of autonomy and control over their own labour. Similarly, the agricultural workers who participated in the Swing riots of the 1830s were not opposed to threshing machines per se but rather to the elimination of winter work, the threat of unemployment, and the overall proletarianization of agricultural labour.

In short, during the first half of the nineteenth century workers were reacting against the encroachment of capitalist social relations, marked by domination and wage slavery, and they were well aware that the introduction of new technologies by their enemies was part of the effort to undo them. Unencumbered by any alien and paralysing notion of technological progress, they simply tried to arrest this assault upon their lives in any way they could. They had nothing against machinery, but they had no undue respect for it either. When choosing between machines and people or, more precisely, between the capitalist's machines and their own lives, they had little problem deciding which came first. As historian Eric Hobsbawm reminds us, unlike their twentieth-century descendants the nineteenth-century "machine breakers were not concerned with technical progress in the abstract." Thus, they were able to perceive the changes in the present tense for what they were, not some inevitable unfolding of destiny but rather the political creation of a system of domination that entailed their undoing. They were also able to act decisively — and not without success when measured in terms of a human lifetime — to defend their livelihoods, freedom, and dignity.

"The machine was not an impersonal achievement to those living through the Industrial Revolution," historian Maxine Berg notes, "it was an issue." In her valuable study of the machinery question in the first half of the nineteenth century, Berg emphasizes that "in the uncertainty of the times, it still seemed possible to halt the process of rapid technological change." Such rapid change, which is in itself destabilizing and thus has been used again and again to force labour onto the defensive, was not at that time viewed as inevitable. Thus, as Berg states, "The working class challenged the beneficence of the machine, first by its own distress then by its relentless protest." It "criticized the rapid and unplanned introduction of new techniques in situations where the immediate result would be technological unemployment." Moreover, "Technological innovation was challenged in everyday struggle in the workplaces of most industries throughout the period. Workers and their trade unions were not ashamed to denounce the type of progress which brought redundancy," speed-up, and loss of freedom. They exposed the reality of the technology, challenged its uses, demanded equitable distribution of gains if there were to be any, and sought greater control over the direction of technological development itself.

"The chief advantage of power looms," the Bolton weavers declared in 1834, "is the facility of executing a quantity of work under more immediate control and management and the prevention of embezzlement and not in the reduced cost of production." The weavers recognized that the power loom "was profitable only for certain fabrics and required a very large investment in fixed capital," as Berg points out. "It was quite clear to many that the productivity of the power loom was not its greatest asset." The weavers recognized that so-called economic viability, the presumed reason for introducing a new technology, was not in reality an economic category but a political and cultural one.

The decision to invest capital in machines that would reinforce the system of domination, which might in the long run render the chosen technique economical, was not itself an economic decision but a political one, with cultural sanction. Other technologies, equally uneconomic but preferable for other reasons, might have been chosen for further investment, and perhaps in the long run they would have been rendered economic. (J.H. Sadler, for example, proposed such an alternative, the pendulum hand-loom, on behalf of the weavers. It was designed to preserve the skills and jobs of the weavers and enable them to avoid the degrading conditions of factory life.)

In short, the weavers raised "a powerful and impressive critique of machinery, a critique that carried a genuine belief that technical change was not a 'given' but could be tempered and directed to match the requirements of social ideals." They "consistently drew attention to piece-rates, home competition, and the specific technical and market conditions for the introduction of power looms." Above all, and again consistently, they demanded a social policy on technology. They proposed, for example, a tax on power looms and a host of other legislative measures to protect the lives of the weavers.

But to achieve their ends the weavers did not rely solely upon such formal tactics. Central to their effort was a strategy of highly organized direct action, the machine-breaking for which they are still remembered today. Between 1811 and 1812, for example, manufacturing workers marching under the banner of the mythical Ned Ludd destroyed over one thousand mills in the Nottingham area. A decade later the machine-breaking spread across the midlands, and as Pierre

Dubois, a historian of industrial sabotage, describes the experience, "In some cases, it had a definitely revolutionary character, involving a confrontation between two armed forces." Workers smashed machines selectively, but deliberately, and this act more than any other characterized the workers' movement of the period.

The precise significance of machine-breaking in the context of the workers' movement is open to interpretation. The paucity of historical evidence makes a fair measure of extrapolation and conjecture inevitable. Most of the revisionist social historians who undertook to reconstruct the movement in the workers' own terms have argued convincingly that the workers were not opposed to the machines per se. They knew who their real (human) enemies were. These historians suggest, therefore, that selective machine-breaking was simply one tactic among others used to cripple and intimidate their foes and win concessions. Rooted in such traditional forms of protest as food riots and incendiarism, machine-breaking also constituted a form of early trade unionism (during a period when such organizations were outlawed) — a form of collective bargaining by riot, as Hobsbawm describes it.

A more recent interpretation, by Geoffrey Bernstein, rejects this minimization of machine-breaking to the status of mere tactic. His analysis suggests that perhaps the social historians are themselves still too bound up in the ideological reverence for technological progress and that to redeem the Luddites as rational the historians somehow had to minimize the centrality of machine-breaking. Bernstein suggests instead that machine-breaking was indeed central, that it constituted a strategy of mobilization for the workers. Such an interpretation appears to be more consistent with the available evidence. All contemporaries — Luddites and those opposed to machine-breaking alike — consistently emphasized that machine-breaking was the hallmark of the movement, its distinguishing characteristic. George Beaumont, a man sympathetic to the workers' plight but opposed to the destruction of machinery, observed that the phrase "I have a good mind to Ned Ludd it" required little explanation at the time.

The Luddites themselves, of course, made no secret of the centrality of machine-breaking and, as Bernstein suggests, "expressed aims tend to be determined by strategic considerations." According to this interpretation, machine-breaking served well to mobilize people with disparate immediate concerns, in different geographical regions, in different trades. It lent a coherence to the movement, encouraging loyalty to a unifying strategy and identification with a few mythical figures (General Ludd and Captain Swing), and it gave the workers a sense of solidarity that magnified their power in their own eyes as well as in those of their contemporaries, including their enemies. Machine-breaking was never the whole of the movement, but it was certainly central, and the success of the strategy is apparent. Rather than isolated acts of resistance soon forgotten, there emerged a movement of great proportions with lasting consequences, a movement still remembered today.

But the way we now evaluate Luddism has not been shaped by the Luddites themselves. Instead we have inherited the views of those who opposed machine-breaking and who succeeded in removing the technology question from the point of production, from the workers themselves, from the present that was the first industrial revolution. In the place of that traumatic reality, these others constructed technological myths about the power of the past and the promise of the future. In the light of these myths the courageous Luddites were made to seem mistaken, pathetic, dangerous, and insane.

"The plight of the workers, made all the more visible by their dramatic protest, shattered the illusion of the beneficence of the emerging capitalist order and discredited once and for all the notion that this society was a realm of shared values and human ends." It is thus not mere coincidence that at this same time society was "discovered" to be a thing apart from the people who comprised it, and that it had a logic of its own that was distinct from and dominated the purposes and aspirations of people. Society as a human artifact, a human endeavour, composed of people, was lost in the wake of capitalism, only to be reinvented as an automatic, self-regulating mechanism in which people were simply "caught up." The hard logic of the market and the machine surfaced supreme, replacing human inspiration, as Lewis Mumford observed, with "the abstractions of constant technological progress and endless pecuniary gain."

Henceforth would "the belief in technological progress, as a good in itself, replace all other conceptions of desirable human destiny." Political notions of justice, fairness, freedom, equality, reason—the hallmarks of enlightened statecraft and the bourgeois revolutions themselves—now gave way to mechanical notions of social betterment. As capitalism revealed its inhuman core, its champions vanished, to be replaced by invisible hands. And social progress became identified with impersonal intermediaries: manufactures, industry, goods, machinery. As human society and people became variables (that is, commodities, factors of production), capital became the constant, not alone the tangible sign of progress but also the imagined engine or cause of progress.

Capitalism, opposed by the workers as a system of domination, exploitation, and alienation, now emerged as simply a system of production identified with progress itself. Such progress, moreover, was viewed as natural and necessary; social prosperity and human happiness would inevitably flow from this automatic process, so long as people allowed it to follow its own natural course, so long as they yielded to the requirements of free competition and untrammelled technological development. If *laissez-faire* became one manifesto of capitalism, *laissez-innover* became the other. "In my opinion, machinery ought to be encouraged to any extent whatsoever," wrote George Beaumont. Ultimately, he believed, such development would fulfil the dreams of the workers because the inventors of machinery were after all the "true benefactors of mankind."

This emergent ideology of technological progress served capitalist development well in the name of material prosperity and diverted attention away from the exploitation entailed. At the same time it shaped all subsequent critiques of capitalism. Even socialists, sworn enemies of capitalist aggrandizement and the profit system, were thereafter compelled to accommodate this new cultural contrivance, to adopt the faith in technological deliverance that had become hegemonic. Indeed, these critics eventually challenged capitalists on the grounds that they alone were the true champions of technological progress and that capitalism merely retarded the development that was possible only under socialism. Thus, a half-century later, Jack London could sum up the socialist creed in a paean to machinery: "Let us not destroy these wonderful machines that produce efficiently and cheaply. Let us control them. Let us profit by their efficiency and cheapness. Let us run them by ourselves. That, gentlemen, is socialism."

Where capitalists maintained that unilinear technological progress, spurred by the competitive spirit and guided by the invisible hand, would usher in a new day of prosperity for all, socialists insisted that such progress would have a double life: moving behind the backs of capitalists, without their knowledge and in defiance of their intentions, the automatic process of technological development would create the conditions for the eclipse of capitalism and the material basis for prosperity under socialism. Both capitalists and their critics, however, had come to worship at the same shrine and, as a result, to reject any opposition to technology in the present tense. How did this happen?

As has already been suggested, such fantasies about technological development arose inescapably as a consequence of the flight from the concrete and the present, which itself reflected the removal of the technology question from the point of production, out of the reach of the unmovable workers. The apologists of capitalism were intent upon fabricating an abstracted worldview that would justify further capitalist development. For them, it was necessary to explain that whatever the all-too-apparent social and human costs of such development in the here and now, social progress was nevertheless being made, with capitalists serving as mere agents of this larger, inevitable, and beneficent process. Political economy emerged to meet this need, largely in response to the workers' actions.

As Berg recounts, "The disruptions caused by mechanization brought in train a legacy of fear," and this led to "the expression of doubts," on the one hand, and "a polemical optimism" on the other, an optimism "based on ignorance." During the first half of the nineteenth century, in the wake of the machine-breaking movement, middle-class apologists and optimistic economists "were missionaries come to spread the gospel of the machine in a land of heretical anti machinery attitudes."

> Middle class economic and political perspectives actively eulogized the progress of science and technology. But, challenged on both sides, by Tory and radical working class opinion, the middle class had to find an explanation for the economic and social impact of the machine. Expressions of wonder at the technical perfection of the machine were not adequate. It was thus that the middle class took to itself a "scientific" theory, political economy. . . . It was not mere coincidence that industrialization and the emergence of political economy occurred at the same time.

The political economists "above all others," Berg insists, were "either optimistic or blind, and possibly both, to the conditions of the working classes." They issued "long and turgid justifications of the introduction of machinery" and insisted upon the ultimate beneficence of technological progress. "Their defense of existing patterns of economic development became in the political setting of the 1830s a strident polemic in favor of capital and machinery," almost a secular religion.

Not all political economists were so easily swept up in such praise of the technological panacea; some, like David Ricardo and John Stuart Mill, recognized full well the legitimacy of the workers' opposition. Thus, in the 1821 edition of his *Principles* Ricardo insisted, "The opinion entertained by the labouring class, that the employment of machinery is frequently detrimental to their interests, is not founded on prejudice and error, but is conformable to the correct principles of political economy." Ricardo was attacked by his colleagues for lending encouragement to the workers' opposition to machinery, but he held his ground. He did, however, support unrestricted innovation out of the fear that if such innovation proved more profitable, foreign competitors would innovate and lure capital out of England, leaving even less employment. For the workers who were displaced, in either case it was in effect a choice of being shot or being hanged, and they remained opposed to the cold logic of competition and the inevitability of technological progress, to not only the machines proper and the machinery of the market, but also to political economy.

In his own *Principles* of 1848, Mill too dismissed as spurious the claims of the apologists of the machine, that machine-building itself would offset the loss of employment caused by machinery or that the introduction of "labour-saving" devices would make work less onerous. "Hitherto it is questionable if all the mechanical inventions yet made have lightened the day's toil of any human being," Mill surmised. Rather, he suggested, "They have enabled a greater population to live the same life of drudgery and imprisonment and an increased number of manufacturers and others to make fortunes," and perhaps they have also "increased the comforts of the middle classes." Nevertheless, Mill insisted upon the ultimate benefit of technological development, not as any panacea but as a means of enlarging the overall wealth of nations.

Thus, even when they recognized the reality of the workers' situation, the economists, as Berg notes, "welded their perception of the advance of technology to their concept of economic development," which proceeded inexorably if not always so benignly through the mechanisms of market, competition, and profit accumulation. But the doctrine of technological progress was not promoted solely in the name of economics. Technological development was also defended in the name of science. The apologetics of capitalism, as Berg suggests, "reached beyond political economy to a far-reaching cultural sphere which took up the machinery question in political economy's terms and made a doctrine of technological progress. This cultural sphere was the scientific movement."

The connection between economically spurred technological development and science, Berg explains, "was promoted both by scientists seeking wider markets for their research and by industrialists seeking some higher rationale for their technological choices and expanding enterprises." In reality, Berg points out, the connection between science and industrial technology hardly existed. "The relationship which was claimed between science and technology was rhetorical only," and, essentially, "The scientific movement of the early nineteenth century acted as a social context for political economy's efforts to demonstrate the benefits of the contemporary industrial transformation." But the cultural connection with science was crucial for the apologists of capitalism. It allowed them to argue that capitalism was a system not only of economic progress but also of science, and that workers who opposed machinery showed not only their selfish contempt for the larger social good but also their ignorance.

And what political economy and the scientific movement failed to do, the true believers in the machine itself, the technical enthusiasts and mystics, accomplished, attributing to machines the force of necessity itself. Thus Charles Babbage, inventor of one of the earliest computers, noted in the 1832 preface to his *Economy of Machinery and Manufactures* that his book was but an application of the principles of his calculating engine to the factory system as a whole, to demonstrate the mathematical precision and predictability that machine-based industry made possible. In the midst of the machine-breaking movement, Babbage contemplated the computer-run factory.

At the same time Andrew Ure, whose description of textile manufacturing served as Karl Marx's point of departure for a critique of modern industry, extolled the virtues of machinery for extending and ensuring total management control over production (as the Luddites well understood). In Ure's mind the factory took on "mystical qualities," as Berg puts it; Ure described the mill as a vast automaton, with all parts in concert, subordinated to the discipline of the self-regulating prime mover, the steam engine. Ure's fantastic vision of the ultimate end of this new discipline, the fully automated factory, like

Babbage's computer-run factory, pictured capitalist industry as the very embodiment of reason, against which worker opposition could not but appear to be futile and irrational. In this view, it was not the fantasists who were the lunatics but the quite realistic and all too level-headed workers who dared to stand in their way.

By the mid-nineteenth century the intellectual dominance of political economy was irrevocably established, and with it the hegemony of apologetics for unrestrained technological progress. A Darwinist view of technological and economic development had evolved that informed state policy and proved relatively immune to the criticisms of both the workers and their supporters. The Tory conservatives who decried the mechanization not only of industry but also of society itself, and who insisted that the social and psychological costs of this progress far outweighed any gains from cheaper commodities, were more easily dismissed as romantic reactionaries. The machine-breakers themselves were assailed with repression and ridicule.

The hegemonic ideology of technological progress, moreover, left its mark on the developing workers' movement as its leaders struggled to be taken seriously in this new intellectual climate. For although they gained strength as a consequence of the workers' actions against machinery, the political champions of labour's cause were no more disposed to follow the workers' lead than were the apologists and agents of capital. They abandoned the workers' strategy not because it proved ineffective but because they believed they knew what was in the workers' best interests, and they were becoming certain that opposition to technological progress was no part of it. Thus the social reformers of the day, whose power in political arenas derived directly from the controversy kindled by the workers' opposition to machinery, acknowledged anyway the inevitability and benefits of technological progress and viewed the workers' plight as the moral problem of poverty, to be solved outside the realm of the economy itself by means of enlightened philanthropy.

The political radicals saw the problem in terms of the distribution of property and political power. They viewed machines simply as tools to be used for good or evil, depending upon who had the power to use them. They decried opposition to machinery as wrongheaded, and they worked to divert workers' attention and antagonism away from the machine and toward the political system. (The workers' critical perspective, as we know, embraced both.)

According to Berg, these efforts to dispel the machinery issue were ultimately successful, and discussion of the machinery question and of the nature and organization of production eventually gave way to discussions of political power and property distribution. "The real grievance," one political radical insisted in 1835, "is neither more nor less than the subjection of the labouring to the monied classes, in consequence of the latter having usurped the exclusive making of the laws. Rents, tithes, taxes, tolls, but above all profits. Here is our distress explained in five words, or to comprise all in one, it lies in the word Robbery.... Machines indeed."

The removal of the struggle from the point of production rendered matters of machinery and production secondary to the political issues that lay beyond the realm of actual production. One result of the political and ideological subordination of the workers by their leaders, then, was a minimization of the matters that the workers themselves initially considered central, and the elimination of the types of direct action that the workers had found to be most effective in their fight against capital. And this diminished debate over and opposition to the introduction of machinery had the effect of ensuring the continued and strengthened hegemony of the doctrine of technological progress, as well as of the capitalist system.

Not all of the champions of labour abandoned the industrial and technological arena. The socialists made it their central battleground. However, they too subordinated the workers to their own peculiar conception of labour's destiny and, in so doing, lost touch with both the concrete and the present. Thus, even though they retained technology as their focus, their perceptions of what technology was and meant became confused and mythological, and tended not only to reflect but also to reinforce the ideology of technological progress. If the capitalists apologized for and rallied behind technological progress, the socialists revered it. For them, technological progress was not simply a means to economic ends and a convenient justification of domination; it was a historical vehicle of emancipation.

The early Owenite socialists viewed the machine in a positive light, as the means of liberation from capitalism and of future prosperity under socialism. They displayed what Berg calls "a wondrous excitement over the machine." Although they saw all too well that, under capitalism and a competitive system, technological innovation led to intensification of work and exploitation, they believed that the same technologies held "something in promise and prospect" in that they could be used to bring about co-operation "in the far time of the

Millennium." The Owenites assumed that technological development under capitalism would lead inevitably to the calamities of overproduction, bankruptcies, and massive unemployment, and that these would so destabilize and weaken capitalist institutions that it would be necessary to abandon competition and private property in favour of a co-operative system and common ownership. At the same time, they believed that the technology would make possible the elimination of the division of labour, and along with it classes, inequality, and domination, and that it would create the material conditions for leisure, education, and collective production in a co-operative socialist society. Thus, on both counts — because technology would undermine capitalism and because it would make co-operative socialism possible — the Owenites condemned anti-machinery sentiment as essentially counter-revolutionary. The Owenite paean to the steam engine, published in the *New Moral World* in 1837, would no doubt have embarrassed even the most strident capitalist apologists: "At length, casting away his guise of terror, this much cursed power revealed itself in its true form and looks to men. What graciousness was in its aspect, what benevolence, what music flowed from its lips: science was heard and the savage hearts of men were melted, the scabs fell from their eyes, a new life thrilled through their veins, their apprehensions were ennobled, and as science spoke, the multitude knelt in love and obedience."

The early socialist's enthusiasm for technological progress was echoed by the so-called scientific socialism of Engels and Marx. In *The Condition of the Working Class in England*, Engels brought together the Tory, Owenite, and political-radical critiques of capitalism. In addition, he introduced the concept of a unified working class, the product of the new machine-based mode of production of industrial capitalism. According to Engels, the new industrial technology, which arose out of the system of competition and exploitation, led inevitably at first to unemployment and the intensification of labour. Thus, in his view, the anti-machinery sentiment of the workers was understandable and justified. However, the new industrial system had also given rise to a coherent industrial working class, with its own organizations and political program of socialism, so that now such proto-unionist and prepolitical sentiments were no longer either necessary or desirable. According to Dubois, the historian of industrial sabotage, Engels believed that "sabotage was the youthful sin of the workers' movement." Now that the movement had become more mature—a direct consequence of technological and industrial progress—such primitive action was counter-revolutionary and had to be opposed.

Engels's colleague Marx took this line of reasoning further, drawing upon the work not only of the Tories, the political radicals, the early socialists, the social reformers, and Engels, but also of the political economists and the philosophers and visionaries of modern manufacturing, Ure and Babbage. For Marx (see the passage from Marx in Appendix II) technological progress was not only the means of capitalist competition, accumulation, and exploitation, but also essential to the advance of modern industry itself—capitalism's contribution to human progress. Modern industry signalled both the transition from hand to machine-based labour and liberation from the drudgery of labour altogether. Technological progress under capitalism was at the same time progress toward socialism, creating the conditions for the demise of capitalism, the living vehicle of revolution (the proletariat), and the material basis for the classless society. Here too technological progress was seen as having a life of its own, with liberating consequences for humanity. To oppose it in the present, therefore, was counter-revolutionary; all those who suffered in the present, in the wake of such progress, were encouraged to accept present technology and look for future deliverance.

By the close of the nineteenth century, then, the ideology of technological progress that had become hegemonic in society as a whole had come to dominate the criticism of that society as well. "Scientific" socialists were quick to disparage and abuse all those who refused to accept technological necessity and acclaim the onward rush of industrial progress, dismissing them as romantic reactionaries or utopian dreamers. Those who continued to uphold the ideas of direct action at the point of production and who opposed the authoritarianism of scientific socialists—those who comprised the left socialists and anarcho-syndicalist tradition—were dismissed as infantile and irresponsible. The Marxists' ridicule of all who opposed capitalist-sponsored technological development thus simply seconded the hegemonic social taboo and further marginalized those who tried to insist upon viewing such development in the present tense.

"The worker will only respect machinery *on the day* when it becomes his friend, shortening his work, rather than as *today*, his enemy, taking away jobs, killing workers." Thus in 1900 Emile Pouget, the French anarcho-syndicalist, echoed the Luddites in defiance of destiny and in the name not of some fabled future but of a pressing present: "Workers have no systematic will to destroy apart from the aim of such destruction. If workers attack machinery, it is not for fear or because they have nothing better to do, but because they are driven by imperious necessity."

But such calls to reason, which surfaced in the syndicalist upsurge in turn-of-the-century Europe and among the followers of the Industrial Workers of the World in the United States, were difficult to sustain in a society now dominated by the romance of technological progress and technological deliverance. Already by the middle of the nineteenth century "progressive workers" like the one portrayed by Elizabeth Gaskell in her novel *Mary Barton* had abandoned such critical reason to become reasonable: "It's true it was a sore time for hand loom weavers when power looms came in. These newfangled things make a man's life a lottery. Yet, I'll never misdoubt that power looms and railways and all such inventions, are the gifts of God. I have lived long enough, too, to see that it is part of his plan to send suffering to bring out a higher good." A half-century later a "disconsolate radical" could lament the fact that "one rarely finds anyone who ventures to deal frankly with the problem of machinery. . . . It appears to infuse a certain fear. Everybody sees that machinery is producing the greatest of all revolutions between classes, but somehow nobody dares to interfere."

Thus had the abstract doctrine of technological progress come to dominate industrial capitalist society. Removed from the concrete and the present, the abstract idea of technological development became simply a given from the past, saturated with the future: autonomous, unilinear, inevitable, and sacrosanct. For both apologist and critic, fatalism and futurism substituted for the present tense; they differed only in their expectations. Meanwhile, the present—where people actually live—was reduced to a mere point in time through which the determining weight of the past and projected flight of the future had momentarily to pass—at best unchallenged and uninterrupted. And this became their legacy, and our inheritance: you can't stand in the way of progress, nor should you—even if it kills you.

Within this profoundly irrational framework, not just the act of opposition to technology in the present but even the mere mention of such opposition became taboo. Indeed, the idea of machine-breaking became more threatening to the ideological edifice than the fact of machine-breaking, which continued without acknowledgement. This taboo was reinforced in the wake of scientific management, which amounted to a new testament of the old gospel, and the rise of science-based industry, which offered progress as its most important product. It was strengthened as well with the further maturation and institutionalization of the labour movement—liberal, social-democratic, and communist alike.

Not that there was no longer any opposition to technology-based changes in working conditions. Such opposition continued and was at times quite dramatic. Yet it remained constrained within the larger ideological reverence for technological progress. And this belief, fuelled by obvious economic expansion and growing abundance, served above all to strengthen the capitalist relations of social domination against which the Luddites struggled. The material prosperity diverted the opposition's attention from the central problem of power—the Luddites' focus—and the fact that capital still had the prerogative to destroy jobs, communities, and lives in the pursuit of profit and in the name of technological progress.

It must be emphasized that this hegemonic ideological inheritance did not rule out opposition to technology in the present on the grounds that it was tactically misguided or strategically shortsighted. Rather, mention of such tactical or strategic possibilities was dismissed without a hearing, and their proponents dismissed as insane. Opposition to technology in the present tense called attention to technology in the present tense, but only for a moment, because the ideology of progress did not admit of such immediacy and fled from it at once, relying not upon evidence or argument but rather upon its power to define the bounds of sanity, of respectable discourse, of reasonable behaviour. The Luddite strategy in the nineteenth century was not debated and found lacking. Rather it was condemned as dangerous and demented, as were all those who identified themselves with it. So too with all latter-day Luddites. To be taken seriously, to be listened to (or even to be heard), one had now to demonstrate allegiance to technological progress, wherever it led. Discussion of present tactics was begged by ideological insistence on this critical point. To violate the taboo was instantly to lose intellectual credibility.

Little wonder, then, that the leaders of labour, who strove so hard to be taken seriously in capitalist society, deferred so readily and totally to this ideology. With regard to technological change, they adopted an official posture of encouragement, accommodation, and acceptance. They were, after all, progressive, and no progressive is against progress. Besides, "You can't stop progress." So, boasting of their maturity and responsibility, they embraced this progress as their own and, in boom times, bellowed of its abundant beneficence.

This is not to say that everyone now actually believed in progress. People still continued to have their doubts about this peculiar and alien notion, and subtly expressed it whenever they talked about such change: "That's progress, I suppose (isn't it?)" "Well, I guess that's progress (isn't it?)" "You can't stand in the way of progress, anyhow (can you?)" The elliptical questions could still be heard, addressed to some absent authority who presumably knew about such things. Yet, even with their barely audible doubts, and even when progress looked pretty grim in the present tense, people were encouraged by social pressure to be respectable, to try to be taken seriously, to look progressive. Those who were not disciplined by their superiors in the ways of progress learned to discipline themselves. For even displaced workers want to be taken seriously and want to make a contribution to society. Thus they must believe that their own sacrifices are suffered for a larger good — how else suffer them with dignity?

And so the Luddites were forgotten, their distant distress recalled only to affirm the primitiveness of their struggle and the insanity of those who dare to repeat it. The term "Luddite" became an epithet, a convenient device for disparaging and isolating the occasional opponent to progress and a charge to be avoided at all costs by thoughtful people. For to be called a Luddite meant that you were not really serious. It meant that you believed you could stop progress. It meant that you were crazy.

It was not that people now knew something the Luddites did not know, nor merely (though this is part of it) that the Luddites knew something long forgotten. Rather, this ideological instinct continued to reflect, and be revitalized by, the sustained political and ideological subordination of people at the point of production by their own friends and leaders. At the same time it reflected those friends' and leaders' own subordination to those who still commanded the rewards and controlled the ideas of society as a whole. It reflected as well their distance from both the concrete and the present. Just how far they had travelled in space and time became abundantly clear once the people at the point of production again began to challenge capital on their own turf, in their own terms, and in the present tense — in the wake of the second industrial revolution.

SOURCES

For the historical account of the first industrial revolution I have borrowed heavily from the works of Eric Hobsbawm, George Rudé, E.P. Thompson, and especially from Maxine Berg, *The Machinery Question and the Making of Political Economy, 1815-1848* (Cambridge: Cambridge University Press, 1980), and Geoffrey Bernstein's provocative unpublished paper, "General Ludd and Captain Swing: Machine Breaking as Tactic and Strategy" (1981). In addition I have used material from Pierre Dubois, *Sabotage in Industry* (Baltimore: Penguin, 1979), and quoted as well from George Beaumont, "The Beggar's Complaint," in *British Labor Struggles* (New York: Amo Press, 1972), and Lewis Mumford, "Technology and Man" (1971).

CHAPTER 2
THE MACHINERY QUESTION REVISITED

Labour's response to the first industrial revolution set a pattern that was repeated in the wake of the second. Once again it was the workers immediately affected by the changes who first sounded the alarm, described the dangers, and undertook direct means to try to slow the assault on their jobs and lives. And once again the issue of technological change was expropriated from the workers by those who spoke for them. The issue was removed from the point of production to executive offices and research centres, where it was fitted into ideological and political agendas of future progress. The result was a loss not just of an understanding of the reality confronting workers but also of a strategy for dealing with it — in the present.

What mechanization was to the first industrial revolution, automation was to the second. The roots of the second industrial revolution lay in the state-sponsored technological developments of World War II. Military technologies — control systems for automatic gunfire, computers for ballistics and A-bomb calculations, microelectronics for proximity fuses, radar, computers, aircraft and missile guidance systems, and a host of sensing and measuring devices — gave rise to not only programmable machinery but also "intelligent" or self-correcting machinery. In the postwar years, the promotion of such technologies was fuelled by Cold War concerns about "national security," the enthusiasm of technical people, management's quest for a solution to its growing labour problems, and by a general cultural offensive to restore confidence in scientific salvation and technological deliverance following the twin traumas of depression and global war. Often with state initiative and subsidy, industrial application of these new technologies (as well as an intensification of older forms of fixed automation and mechanization) began to take hold, in steel, auto, petroleum refining, chemical processing (and uranium enrichment), and aircraft, machinery, and electrical equipment manufacture, among others.

The threat to established work rules, working conditions, and job security posed by the introduction of such technological changes sparked strikes, sporadic sabotage, and, during the late 1950s and early 1960s, a wide-ranging debate about the social implications of automation. The trials of the longshoremen facing containerization, the printers facing teletypesetting and computers, and refinery workers confronting computer-based centralized process-control were the focus of attention. Despite the efforts of rank-and-file workers in these industries to prevent or at least slow down the introduction of these technologies (which had been designed, in part, to reduce their power as well as numbers) through the use of strikes and other forms of direct action (as well as demands for veto power over the decision to introduce the new systems, as proposed by the printers), their unions uniformly bowed to the hegemonic ideology of progress. Denying steadfastly that they were against technology, union leaders strove to avoid media charges of Luddism and either conceded the futility of opposition and yielded, or endorsed the notion that such technological changes were the surest route to prosperity.

Meanwhile, union leaders used the same charges of Luddism against more militant union members who refused to comply. While some unions did succeed in gaining a measure of compensation and job protection for some of their members, they all yielded completely — over significant rank-and-file protest — to management's exclusive right to decide on new technology. By 1965, government spokespersons were offering reassurances that fears about automation were unwarranted. These reassurances seemed credible in a period of economic expansion, even though that expansion was largely inspired by the state-sponsored enlargement of the "service sector" and the promise of a Vietnam War boom. Economists revealed that the very idea of technological unemployment was simply a semantic confusion, since technological development invariably created more jobs than it eliminated. In reality, the costs of the changes were concealed in the aggregate by an expanding economy and the temporary absorptive power of the service sector, as well as by the fact that in most cases the new technologies were still in their infancy and their full impact was yet to be felt. But the union leadership prematurely abandoned the struggle and, echoing the official optimism, reaffirmed the ultimate beneficence of technological progress.

Those who continued to lose their livelihoods, or watched the steady deterioration of their working conditions and the erosion of their control over their jobs, were undoubtedly not amused, but neither were they heard. Their plight remained hidden, despite frequent subdued rumblings, while the nascent technology of automation began to reach maturity and find fuller and wider application. Thus, the increasing displacement, deskilling, and disciplining of workers in industry proceeded apace, largely unnoticed except by the workers themselves until, by the end of the 1960s, the situation exploded in an upsurge of pent-up rank-and-file militancy.

The late 1960s and early 1970s were marked by an outpouring of worker initiative, cynicism, and rage about union leadership collaboration, and a renewed emphasis upon direct action. In June 1970, *The New York Times* noted "signs of restlessness in the plants of industrial America, and increasing dissatisfaction and militancy." "At the heart of the new mood," the *Times* observed, "there is a challenge to management's authority to run its plants," as well as a distrust of traditional union and collective bargaining processes: "The older generation would have filed a grievance. The young people have no faith in that. They want it settled right away." There is a feeling of "not wanting to be an IBM number, just part of the machinery," the *Times* added, concluding, "There's a big explosion coming in the industrial unions." The *Wall Street Journal* noted the same month that the number of grievances had grown enormously, primarily against layoffs, the downgrading of workers, and speed-ups. According to the *Journal*, many experienced observers considered the situation "the worst within memory."

Workers such as the teamsters, post office and telephone employees, coal miners, and steelworkers resorted to slowdowns, absenteeism, wildcat strikes, and other means of direct struggle, to the extent that *Fortune* had to alert its readers that management was "dealing with a workforce . . . no longer under union discipline." In addition to these traditional forms of protest, there arose new forms of direct action, shop-floor organization, counter-planning strategies against management, rank-and-file caucuses against union leadership, and systematic sabotage. In an automobile-engine plant in Detroit, for example, one worker observed "plant-wide rotating sabotage programs." Workers there took turns sabotaging the production process, routinely producing damaged or defectively assembled products until reject rates of 75 per cent forced unscheduled shutdowns of the entire plant. "The biggest issue really comes down to what we working people are going to accept as 'progress,'" declared a leaflet put out by the Longshoremen of San Francisco during their strike of 1971. "We, like many other workers, are faced with a technological revolution of new 'labor saving' devices and methods of operation. This is what our employer means by 'progress.' But, if this 'progress' is left unchecked it will simply mean that our employer will line up at the bank with ever bigger profits, while we line up at the unemployment and welfare office. It is essential for labor," these workers insisted, in defiance of both management and their own international, "to challenge the notion that the employer — in the name of 'progress' — can simply go ahead and slash his workforce or close his factory or, as is being planned in our industry, close an entire port, and to do this without any regard for the people and community involved."

This upsurge in rank-and-file direct action was symbolized by the wildcat strikes and creative sabotage of production at Lordstown, General Motors' most automated assembly plant. There workers openly defied the new production system, and according to Pierre Dubois, their actions "freed" them "from inhibitions and their fear of management." In addition, the actions liberated them from traditional taboos and the mindset of technological determinism. In their protest against degraded working conditions, they proclaimed that technological progress was a political rather than an automatic, inevitable, natural process. Thus their protest gave rise to a radical critique of the neutrality of science and technology. Moreover, as Dubois observed, they "had the satisfaction of having dared to maltreat their equipment." Such direct action at the point of production was by no means limited to the United States; it erupted with equal or greater force in France, Italy, Germany, Scandinavia, and elsewhere. There too it signalled a rise of rank-and-file power within the unions and thus a return to the challenge of the second industrial revolution in the concrete and in the present.

Dubois observed in 1975:

> The spread of sabotage indicates a determination by the rank and file to make their own decisions. The waves of strikes in the late 1960s resulted in giving the rank and file far more power [and] the spread of various forms of sabotage . . . is due to workers on the shopfloor having a greater say over what forms of struggle they will adopt. The fact that some unions now leave their membership free in this respect is directly due to shopfloor pressure. People can decide on the most appropriate form of action in relation to their own particular situation, and may well decide on sabotage if it seems advantageous. It is largely the increasing role played by workers as a whole in running their own struggles that explains the increase in sabotage.

As workers fought to overcome their political subordination within their own organizations, their plight became more visible. And as they began to achieve their aim, the struggle against capital became less ideologically constrained and thus more direct and effective. As workers became sophisticated about the ways in which the new integrated automation systems rendered management even more vulnerable to sabotage than before, new forms of direct action emerged and spread throughout the workforce, to the skilled and unskilled alike, young and old, unionized and non-unionized, men and women, veterans and new industrial workers, in all industries.

In searching for new forms of struggle, and encouraged by an expansive economy, workers regained their confidence and asserted their own power — over technology, progress, historical necessity, capital, management, expertise, and union leadership. Acting upon their own experience they took their fate into their own hands and, for a moment at least, shook the world. As historian Jeremy Brecher wrote at the time: "Today, there is an enormous cynicism about leaders and organizations of all sorts. This cynicism often looks like apathy, especially to aspiring leadership groups like various insurgents and leftist parties. . . . But it also means that if and when large numbers of workers again move into action, they will be better inoculated against the appeals of 'leaders' and may try to keep control of the struggle in their own hands." Brecher was correct, and he was not alone with this understanding. Almost immediately, unions, managements, and governments, recognizing the danger to the established institutions, strove to recapture the initiative, regain control, and quiet the rebellious ranks. Among other things, this effort once again entailed removing the issue of technology from the shop floor, from the workers, and from the present.

As with the Luddite revolts of the first industrial revolution, resistance to the second industrial revolution was met with repression. People were disciplined, jailed, isolated, and otherwise intimidated. In 1970, for example, France passed a new law against "all instigators, organizers, or deliberate participants in sabotage." While using the upsurge to advantage at the bargaining table, the unions—liberal, social-democratic, and communist alike—condemned much of the direct action and publicly distanced themselves from it.

Management responded to the wave of rank-and-file militancy with disciplinary measures, lockouts, and legal devices, as well as by designing and introducing new technology that, it was hoped, would diminish the possibility of worker intervention in production or eliminate the need for workers altogether. In addition to these traditional responses, the managers of some companies experimented with new methods—so-called job enrichment, job enlargement, and quality of worklife schemes—designed to absorb discontent and redirect energies along more productive paths. Sweden was a centre for such experimentation and became a model throughout the industrialized world.

"Far from being motivated by a new cooperative attitude between labor and capital in Sweden," however, as auto-industry historian David Gartmann notes, "these changes in technology and work [were] the results of renewed class struggle." Sweden too had been struck by an epidemic of worker rebellion and resistance even more severe than in the United States; absenteeism, labour turnover, and wildcat strikes had escalated dramatically. At Volvo, daily absenteeism had reached over 15 per cent and annual labour turnover peaked at over 50 per cent. "The main reason for such resistance," Gartmann states, was "discontent with the stultifying, monotonous, and intense nature of the work itself," which was reinforced by the introduction of automation. P.G. Gyllenhammar, Volvo president at the time, acknowledged, "Labor unrest that became visible in 1969 made it necessary to adapt production control to changing attitudes in the work force."

In the United States, many companies also initiated job-enrichment schemes to try to regain the loyalty and co-operation of the workforce as well as to ensure the fullest utilization of expensive new equipment. Most of these experiments succeeded in terms of increasing productivity, output, and quality, and reducing absenteeism and turnover, but they were terminated once the workforce began to use its expanded responsibilities to try to extend further its control over production. Nevertheless, for a time at least, the so-called humanistic approach to management held sway and became official gospel with the publication in 1973 of *Work in America*. This study was the product of a special task force commissioned by HEW Secretary Elliott Richardson in 1971 in response to the emergence of what he condescendingly referred to as the "blue collar blues" and the "white collar woes."

"Great care must be taken to interpret wisely the signs of discontent among workers," the commission advised. "Increased industrial sabotage and sudden wildcat strikes, like the one at Lordstown, portend something more fundamental than the desire for more money.... The impact of technology has been acutely felt by the blue-collar worker, not necessarily because it puts him out of a job but because it lowers his status and satisfaction from the job.... While many industrial engineers feel that gains in productivity will come about mainly through the introduction of new technology," the commission pointed out—and this was the attitude behind the design of Lordstown—the result was collective resistance, wildcat strikes over the pace of production, and opposition to "robot-like tasks." In general it gave rise to a growing hostility to the traditional forms of management, the "anachronistic authoritarianism of the workplace." Thus, in the wake of the second industrial revolution, the commission found that "the productivity of the worker is low—as measured by absenteeism, turnover rates, wildcat strikes, sabotage, poor quality products and a reluctance of workers to commit themselves to their work tasks."

The commission was concerned that such dissatisfaction could have serious consequences beyond the workplace as well as within it. If industry was suffering from low productivity and high rates of sabotage, absenteeism, and turnover, the unions were paying a price through "the faltering loyalty of a young membership that is increasingly concerned about the apparent disinterest of its leadership in problems of job satisfaction." For the nation as a whole, the commission warned, the discontent was resulting in staggering health bills and mounting crime and delinquency. "Most important" were the political consequences: "the discontent of women, minorities, blue-collar workers, youth and older adults would be considerably less were these Americans to have an active voice in the decisions in the workplace that most directly affect their lives."

To stem this growing threat of rebellion, increase productivity, and reinforce the stability of existing institutions, the commission proposed greater "participation": "several dozen well-documented experiments show that productivity increases and social problems decrease when workers participate in the work decisions affecting their lives." The commission thus reflected and reinforced the effort to substitute participation for power, experiments orchestrated from above for resistance and "counter-planning" organized from below.

Throughout the country, many unions entered into such co-operative participation relationships with management, with government encouragement, and, as a result, enlarged the scope of bargaining and their involvement in management activities. For the unions, then, the upsurge in rank-and-file militancy led to some victories vis-à-vis management. But for the workers themselves, whose lives were temporarily enlarged as a result, such victories proved short-lived and severely limited. More important, in the wake of these limited gains, the rebellious energies that had brought them about dissipated and all but disappeared. In their place arose committees, rules, agreements, and other formal devices for dealing with the new challenges at the workplace, including the challenge of new technology. While the upsurge of rank-and-file militancy revived interest in the workplace among social critics and gave rise to a penetrating critique of the modern science and technology of production, the defensive response to that upsurge gave rise also to a formalization of the technology issue. As critics re-examined the social, political, and cultural dimensions of "progress," corporations, governments, and trade unions (especially in Europe) established new bureaus, programs, research projects, and centres intended to co-opt the rebellion by institutionalizing it. It worked. Before long, new academic disciplines in "technology and work" had gained respectability, generating a new form of professional career. The new professionals were called "technology researchers" in Scandinavia and elsewhere; the more politically motivated became known as proponents of "action research." Whatever their motivation, sympathies, political commitments, or intellectual interests, they all made a career of the problem above and apart from (and in some cases, substituted for) the practical challenges facing workers. As a result of this vested interest, and regardless of their other purposes, they sustained and pushed forward the formalization of the technology issue.

Still building upon the energies unleashed by the workers themselves, the professionals soon produced a plethora of publications, conferences, and research, and assisted the trade unions in formulating new contract language and, ultimately, new agreements on the introduction of new technology. Whatever these gains, however, they were achieved at the expense of removing the technology issue from the shop floor and thus from the realm of direct action available to the workers themselves. "With increasing formalization," Dubois observed, "the spread of sabotage could once again be held in check by pressure from trade union organizations opposed to it." At the same time, as a consequence of its removal from the point of production, discussions of the technology question became increasingly abstract and future-oriented, abandoning the present as a realm of the struggle.

The so-called new technology agreements originated in Norway, the product of an unprecedented "action-research" collaboration between the Iron and Metal Workers Union and the Norwegian Computing Centre. Launched in 1969 as a direct result of the rank-and-file upsurge, the Norwegian project gave rise to both local and national agreements over the introduction of new technology, "data" shop stewards (specifically charged with policing the new technology), and formal trade union participation in decisions about the introduction and use of new technologies. The agreements, grounded upon social-democratic legislation for protection of the work environment, in principle went beyond previous agreements over technology such as those pioneered by the Longshoremen's and Warehousemen's Union and the International Typographical Union in the United States. They were not confined to the post hoc protection of workers from the consequences of progress but were intended to give unions a say in what the progress itself would look like. Following upon the pioneering Norwegian experience, similar projects and agreements were formulated in Sweden and Denmark while, around the same time, parallel formal processes were established in Holland, West Germany, and Italy—all, again, in response to the general climate created by the wave of rank-and-file militancy. In principle, these agreements constituted a significant advance for the trade unions and a potential challenge to traditional management prerogatives. In practice, however, they were rarely used to prevent the introduction of new technologies. Indeed, the agreements often served to circumvent worker opposition to the introduction of new technology.

In Denmark, for example, the central federation of trade unions successfully opposed widespread worker insistence on the right to veto new technology. In England, Barry Wilkinson found that agreements were reached only after the technological changes in question had been implemented, and that they remained concerned solely with post hoc matters of payment, redundancy, and retraining. Wilkinson concluded from his study of the politics of new technology that "traditional methods of bargaining are wholly inadequate for technological change" and, more important, that "despite the current popularity of new technology agreements" and "the flurry of publications on 'the new technology' . . . bargaining over skills and the organization of work remains at an unofficial, often covert level." In other words, the real struggle over the new technology has continued to take place on the shop floor itself (in outright resistance, ad hoc negotiations, and sabotage), regardless of and sometimes in spite of the formal agreements. Workers have increasingly learned not to rely too heavily upon formal agreements for the protection of their jobs and working conditions.

While the new technology agreements have perhaps provided some post hoc protection for workers, they have had little or no impact upon the actual design and implementation of technological changes. Indeed, it has been suggested that, if anything, the agreements have probably facilitated the introduction of the Trojan horse of new technology within the shops, that the trade unions with their formal agreements have—in the words of Stan Weir, one rank-and-file opponent of the ILWU "Mechanization and Modernization" Agreement of 1960—"run interference for the new technology" by weakening workers' resolve and ability to resist. Removed from the shop floor, the issue of technology has been formalized and packaged and then returned, from above, in a form that generates both false security and confusion and eliminates the possibility of direct action. Thus, the formal agreements, while in principle signifying a challenge to established management rights, in practice have perhaps taken the teeth out of such a challenge. And, despite their formal stance against the harsh consequences of technological change, trade union leaders have continued to echo the proud pronouncements of the past that they are not, after all, against progress.

At the same time, reacting to the accelerating technological agenda of management—which has always used change as a tactic to disorient its opposition—the unions have been forced onto the defensive. Trying to hold on and keep track of (if not pace with) new developments, the unions have been forced to focus upon what is changing (technology) and to ignore what is not (the dominant relations of power). While this frenetic exercise in futility has done little to help the unions and their members find a way out of their predicament, it has provided a great deal of full-time work for researchers. Seduced into the details of the technology and endlessly documenting the horrors, they have intensified the trade union obsession with professional rather than worker competence and even lent a degree of polite respectability to the unions' efforts. Most important, they have reinforced a fundamental confusion about the social realities of technological development.

The recognition that technology is political constituted an important ideological breakthrough since it overcame the fatalism of technological determinism, long a staple of capitalist apologetics. But there are at least two possible conclusions that could be drawn from this belated insight. First, the understanding that technology reflects power relations in society could imply that those with more power would continue to determine the shape and direction of technology for the foreseeable future. Therefore, the conclusion to be drawn would be twofold: in the long run to try to shift the balance of power, and in the short run to do everything possible to prevent the introduction of the present technology, since it reflects the interests of those in command. Those few who have experimented with this position have invariably stumbled upon the taboos against Luddism, the cultural compulsions of progress, and economic deterministic arguments about efficiency, productivity, and competitiveness. Thus, they have always opted for a formalistic approach and settled for bargaining over technology post hoc and from a position of weakness. There has been little evidence of any unions actually mobilizing workers to try to increase their power vis-à-vis management and even less of any concerted attempt to organize opposition to the introduction of new technology.

The second inference from the insight that technology is political is seemingly less sobering and more liberating. Since politics is the art of the possible, and technology is understood to be political, with technology too, then, anything is possible. This more optimistic interpretation has generated a great deal of enthusiasm about possibilities and led to a fetish for and fantasies about alternatives. At the core of this interpretation, it must be emphasized, was an important advance beyond the technological determinism of both liberal and Marxist notions of technological progress. It signalled a rejection of the perception of technological development as unilinear, inevitable, and automatic, and a recognition that political and social change would require a change in the forms of scientific and technological theory and practice. This new, expansive view of technology offered hope of transcending the mere defensive posture of labour. Rather than reacting endlessly, off balance, to management's technological agenda, labour could now go on the offensive itself by formulating its own alternative technological agenda.

But this insight about the wide range of technological possibilities, only narrowly reflected in the existing social context, gave rise inevitably to confusion and false promises. Some proponents of this interpretation of technological politics assumed that alternatives could be created without a change in power and even that such alternatives would of themselves bring about a change of power. This way of thinking, most common among technical people still imbued with ideas of technological deliverance (and hoping to change things without having to change careers), followed from a logical fallacy: technology reflects politics, therefore change the technology and this will change the politics. In essence, this new (old) habit merely reintroduced technological determinism in a different form.

More sophisticated proponents of this interpretation of technology as politics, while less enthusiastic about technological solutions per se, nevertheless have emphasized the need to develop independent knowledge and competence about technological possibilities. In this view, the effort to develop alternatives gives confidence and direction to an otherwise defeatist and fatalistic cadre. As an organizing device, moreover, it is held that alternatives serve to inspire, embolden, raise consciousness about political realities, and provide something to fight for rather than merely against, something to believe in. Finally, the proponents contend, correctly, that while alternatives are possible in theory, for the most part they do not exist and must somehow be invented. Only then will the possibilities be rendered concrete, a visible demonstration of another route of progress.

But this interpretation of technology as politics has ignored some basic realities. The existing technologies reflect centuries of continuous development along a particular path, and the development of alternatives will similarly require years of reflection, research, and practical experimentation. It will not be possible to turn around the legacy of the industrial and scientific revolutions overnight. Such fundamental changes are a vital political task, but what is to be done *now*? What good is a strategy for the future without a strategy for survival in the present? Even if the unions devoted all available resources to the development of alternatives, it would still be years before anything emerged reflecting labour's interests. Moreover, at present, no unions have sufficient power at the bargaining table or anywhere else to demand and enforce a fundamental redirection of technological development and, thus far, organizing efforts reflecting this approach have aroused little interest among workers. Even if unions devoted all available resources to this organizing effort, and succeeded, it would still be years before they could marshal the power sufficient to influence the shape of technology. Meanwhile the present technology continues to enter the shops and erode the potential of labour power. Will the unions survive long enough to be able to redirect the development of technology?

The Luddites had some effect in slowing down the advance of the first industrial revolution and thereby bought themselves some time, some space — precisely what the unions now have so little of. But the Luddite effort entailed a massive insurrection — organized by the workers themselves and including direct opposition to technology in the present tense — that took the British army decades to quell completely. Thus far, few unions have given any indication that they imagine the need for, much less that they are prepared for, such a struggle.

The appeal to alternatives thus diverts attention from the realities of power and technological development, holds out facile and false promises, and reinforces the cultural fetish for technological transcendence. In short, having overcome the ideology of technological determinism, the fatalism of the past, it flips immediately into fantasies of the future. Not only does this reinforce the hegemonic ideology of technological progress, but it still leaves the present essentially untouched.

Throughout the industrialized world, unions have succumbed to this tendency, abandoning the present in quest of a different future. In Norway, Denmark, and Sweden, where the new technology agreements first took hold, there has been a notable shift in this direction. In Norway, computer scientists have secured the support of unions to develop long-range labour-oriented research projects on the "office of the future" and the "shop of the future." In Sweden, one union has embarked upon a project to redesign video-display terminals. Researchers and unions throughout Europe have been contemplating the joint development of a union-controlled communications computer net. Action research veterans in Denmark and Sweden have embarked upon a long-term effort with unions and manufacturers to develop a worker-friendly computer-based printing system, in a project appropriately entitled UTOPIA. As Anders Hingel of the EEC, a long-time consultant on technology to Scandinavian trade unions, notes, "There exists a definite inclination to present alternatives to the development of the laissez-faire technology."

In the United States the International Association of Machinists (IAM) has recently formulated a "technology bill of rights" program that belatedly embraces the European "technology agreement" strategy as well as the subsequent enthusiasm for alternatives. Explicitly modelled after the experience of the Norwegian Iron and Metal Workers, the IAM program is a perfect example of the latest, most sophisticated trade union response to the challenge of new technology. The union is not against progress; it just wants to participate in the decision-making and thereby steer progress in a more humane direction. We "seek full participation in the decisions that govern the design, deployment, and use of new technology," IAM president William Winpisinger explained to a Congressional subcommittee. "The objective," the IAM emphasizes, "is not to block the new technology but to control its rate and the manner of its introduction, in order that it is adapted to labour's needs and serves people, rather than being servile to it or its victim." The union, reflecting the fertile faith in alternatives, has insisted that "It can go either way," but "it's headed the wrong way now."

Recently the union has embarked upon a campaign to get the message down to the membership, while at the same time it has collaborated with technical people from universities in an effort to develop its own technical and managerial competence, to try to prepare itself for its new, innovative role. It has also formulated a set of basic demands to be used in future contract negotiations and formal agreements. But meanwhile, design and deployment of the new technology have continued without IAM participation, enabling management to drastically reduce union ranks and power. Thus far, there has been no indication that the union is doing anything in a practical and immediate way, or in a way accessible to workers on the shop floor, to try to stop this technological assault. While the preamble to the IAM Technology Bill of Rights declares that "uses of technology that violate the rights of workers and the society will be opposed," there is no indication that this is anything more than rhetoric. There is no hint as to what form such opposition might take. Thus, while rhetorically a challenge to management rights and a bold initiative in a new direction, the approach offers merely more of the same: the appearance of struggle without its substance, allowing unions to bemoan "progress" without actually having to stand in its way.

Whether or not this union strategy serves the interests of workers in the present, it does preserve the progressive respectability of union leaders and provides a veritable field day for researchers. The prospects for futuristic research are, after all, infinite. Also, trying to develop technically and economically viable alternatives is a never-ending enterprise insofar as it ignores the fact that these are not really technical or economic categories at all, but political and cultural ones. No existing technologies have ever had to pass such tests of viability until (if ever) after the politically determined and culturally sanctioned decisions to invest in them had already been made, on other grounds. Thus the effort to render new alternatives realistic in economic or technical terms, already under way in several projects, is a Sisyphean task, consuming scarce resources and likely to end in frustration and cynicism. For without the requisite social power that could deem labour's alternatives viable—whether economic in present terms or not—whatever the researchers and unions come up with will be dismissed on economic and technical grounds, but for political reasons. Nevertheless, the research proceeds apace.

Roy Moore and Hugo Levie of Ruskin College, for example, have worked on a project on "the impact of new technology on trade unions in England." They recognize that "a struggle for some control over technological change and any related work organization will, in the coming period, be one of the most important tasks for trade unions," and they warn that, if unions fail to heed this challenge, "the penalties will be high in terms of unemployment and social dislocation as well as industrial relations disruptions" (initiated by rank-and-file workers). To avoid these disasters, they argue, the unions must, above all, invest in more research.

Although they concede that their studies to date "have inevitably fallen short of actually helping the trade union representatives to influence technological change," they still call for "further, longer-term research on a wider basis" and point out, "It will cost money — a lot of money." Unabashedly unreflective about their own interests in sustaining such an expensive enterprise, the two researchers insist that this is the key to union and worker salvation in the wake of the second industrial revolution. Meanwhile, offered such costly future-oriented strategies and confronted in the present by an intensifying technological assault, at least some workers are beginning once again to have their doubts, and to take their fate into their own hands.

SOURCES

> *The New York Times*, June 1, 1970, July 21, 1971; *Wall Street Journal*, June 26, 1970, November 20, 1970; Thomas O'Hanion, "Anarchy Threatens the Kingdom of Coal," *Fortune*, January 1971; Pierre Dubois, *Sabotage in Industry* (London: Penguin Books, 1973); Jeremy Brecher, *Strike!* (Boston: South End Press, 1972); David Gartmann, "Basic and Surplus Control in Capitalist Machinery," *Research in Political Economy* 5 (1982); Special Task Force, *Work in America* (Cambridge: Massachusetts Institute of Technology Press, 1973); Barry Wilkinson, *The Shopfloor Politics of New Technology* (London: Heinemann Educational Books, 1983); William W. Winpisinger, "Written Testimony: Skilled Manpower and the Rebuilding of America," presentation to the Subcommittee on Economic Stabilization of the Committee on Banking, Finance, and Urban Affairs, U.S. House of Representatives, July 24, 1981; Dick Greenwood, "Starvin' in Paradise with the New Technology," unpublished paper, 1982; Roy Moore and Hugo Levie, "New Technology and the Trade Unions," *Workers' Control* No.2 (1982); P.G. Gyllenhammar, *People at Work* (Reading, Mass.: Addison Wesley, 1977); Bill Watson, "Counter-planning on the Shop Floor," *Radical America*, May-June 1971; Henning Tjornehoj, Danish LO research director, lecture, Technical University of Denmark, December 1982; Anders Hingel, "Survey of Scandinavian Trade Unions' Response to New Technology," unpublished EEC report, 1982; David L. Goodman, "Labor, Technology, and the Ideology of Progress: The Case of the New York Typographical Union Local 6," B.A. thesis, Harvard University, 1983; ILWU Local 10, "Longshore Strike," strike leaflet, San Francisco, August 1971 (courtesy of Herb Mills).

CHAPTER 3

PRESENT-TENSE TECHNOLOGY

"Everyone believes the United States is in the midst of an economic transformation on the order of the Industrial Revolution," *Business Week* noted in the early 1980s. But this fashionable analogy between today's second and yesterday's first industrial revolution is only half complete: the catastrophe has been left out. The prospect of another epoch-making historical leap thus generates simple-minded delight among those managers who seek to enlarge their authority at the expense of workers; among those equipment vendors whose high-tech hype enchants the unsuspecting; among those man-child technical enthusiasts who are encouraged to indulge their socially irresponsible fantasies at public expense; among those system-building militarists who imagine security through strength through silicon; among those trade unionists who remain handicapped by the hallucinogenic homilies of technological progress; and among those ambitiously neoprogressive politicians whose rosy rhetoric belies their ignorance of that first "great transformation"—"a world turned upside down," contemporaries soberly described it—and the mass insurrection that followed in its wake. For a more complete analogy would shake the spirit, not stir it, and give thoughtful people pause. It has been forgotten in the present paeans to progress that the earlier episode was stained in blood as well as grease and that it engendered not only passive immiseration but also active rebellion.

"To enter into any detail of the riots would be superfluous," Lord Byron told his colleagues in the House of Lords in 1812 during the height of the Luddite uprising. "The House is already aware that every outrage . . . has been perpetrated, and that the proprietors of the frames [textile machinery] obnoxious to the rioters, and all persons supposed to be connected with them, have been liable to insult and violence." He continued:

> During the short time I recently passed in Nottinghamshire, not twelve hours elapsed without some fresh act of violence; and on the day I left the county I was informed that forty frames had been broken the previous evening, as usual, without resistance and without detection.
>
> Such was then the state of that county, and such I have reason to believe it to be at this moment. But whilst these outrages must be admitted to exist to an alarming extent, it cannot be denied that they have arisen from circumstances of the most unparalleled distress: the perseverance of these miserable men in their proceedings tends to prove that nothing but absolute want could have driven a large, and once honest and industrious, body of the people into the commission of excesses so hazardous to themselves, their families, and the community.
>
> At the time to which I allude, the town and county were burdened with large detachments of the military; the police was in motion, the magistrates assembled; yet all the movements, civil and military, had led to—nothing. Not a single instance had occurred of the apprehension of any real delinquent actually taken in the fact, against whom there existed legal evidence sufficient for conviction. But the police, however useless, were by no means idle: several notorious delinquents had been detected—men, liable to conviction, on the clearest evidence, of the capital crime of poverty; men, who had been nefariously guilty of lawfully begetting several children, whom, thanks to the times! they were unable to maintain.
>
> Considerable injury has been done to the proprietors of the improved frames. These machines were to them the advantage, inasmuch as they superseded the necessity of employing a number of workmen, who were left in consequence to starve. . . . The rejected workmen, in the blindness of their ignorance, instead of rejoicing at these improvements in arts so beneficial to mankind, conceived themselves to be sacrificed to improvements in mechanism. In the foolishness of their hearts they imagined that the maintenance and

well-doing of the industrious poor were objects of greater consequence than the enrichment of a few individuals by any improvement, in the implements of the trade, which threw the workmen out of employment and rendered the labourer unworthy of his hire.

And it must be confessed that although the adoption of the enlarged machinery in that state of our commerce which the country once boasted might have been beneficial to the master without being detrimental to the servant; yet, in the present situation of our manufactures, rotting in warehouses, without a prospect of exportation, with the demand for work and workmen equally diminished, frames of this description tend materially to aggravate the distress and discontent of the disappointed sufferers. . . .

These men never destroyed their looms till they were become useless, worse than useless; till they were become actual impediments to their exertions in obtaining their daily bread. . . . These men were willing to dig, but the spade was in other hands; they were not ashamed to beg, but there was none to relieve them; their own means of subsistence were cut off, all other employments preoccupied; and their excesses, however to be deplored and condemned, can hardly be subject of surprise.

One hundred and sixty-three years after Lord Byron's observation, people were still being caught by surprise. In the predawn hours of October 1, 1975, the lone foreman in the pressroom of the *Washington Post* was jumped by several desperate and determined men, one of whom held a screwdriver to his throat. Helpless, the foreman watched silently while, for the next twenty minutes, a team of highly skilled pressmen — whose jobs were threatened by the introduction of computerized "cold type" technology — systematically disabled all seventy-two units of the *Post*'s nine presses. Methodically, they sliced through the cushions on the press cylinders, ripped out the electrical wiring, removed the detachable chucks required to support the one-and-a-half-ton reels of newsprint, cut air hoses, and damaged scores of reels. The printing plates were locked in place and the locking keys, inserted in holes on the press cylinders, were broken off and the cylinders jammed. The most serious damage was then inflicted by fire, but only after the automatic fire extinguisher had been duly disabled.

Later that morning the president of the local pressmen's union refused to accept blame for the damage, insisting that his members had been "frustrated" by management intransigence and, as a consequence, "just went crazy and panicked" in what he described as a moment of "temporary insanity." But the *Post* management knew better. For all their surprise, a *Post* spokesman acknowledged that the attack appeared "to have been executed by people who had pre-planned and synchronized their actions. It would be impossible for these kinds of damage to be done in that short a time without a plan, without assigned tasks, and without people who knew precisely what they were doing."

This extraordinary event at the *Washington Post* received considerable media attention for a while *(Time* dubbed the pressmen "the Washington Luddites"), but was too quickly forgotten: the pressmen, the familiar story went, had been rendered obsolete by the new technology; given to "excesses" by the futility of their plight (they were ultimately replaced at the *Post*, their strike broken, and their union destroyed), their extreme action was but a final gesture in a tragic story with an inevitable ending. But it is perhaps more likely that the real significance of this episode belongs to the future rather than to the past, that it signalled not an end but a beginning. The pressmen may have been not behind their times but ahead of them.

The Luddites, it will be remembered, were not against technology per se. They were contending with the social relations of industrial capitalism and the increasing dominance of the "economy"—and of those who dominated the economy—over society. Society was being reduced in both theory and practice to a mere reservoir of factors of production for enterprise: not only the land and the trees were becoming commodities, but people too, all to be used and disposed of as economic expediency required, as judged by the cold calculus of accumulation. The introduction of machinery was but one rather visible and tangible manifestation of this social upheaval, one that reflected the extension of capitalist control beyond commerce to the activities of production itself. Marx was not writing figuratively but literally when he decried the debasement of human beings to mere appendages of machinery and veritable slaves of those who owned capital. And today, when respectable discourse still requires euphemistic substitutes for "capitalism," it is difficult to remember that this term was itself a euphemism of sorts, a polite and dignified substitute for greed, extortion, coercion, domination, exploitation, plunder, war, and murder. This was the list of grievances compiled by the Luddites in their heroic defence of society. Machine-breaking was simply a strategy and a tactic for correcting these violations of morality and humanity, violations that were later obscured by myths of the market and technological progress.

Today, people are once again having to contend with a major change in social relations, this time occasioned by the multinational extension of corporate capitalism, operating in a global labour market. Once again, the transformation is being facilitated and reinforced (and obscured) by the introduction of new technology: computer-based communication and production systems. These latest devices give capital a new mobility, enabling capitalists to pick and choose from the world's reservoir: societies and peoples played off against one another in search of the cheapest and most servile hands. Moreover, these new technical systems hold out the prospect not simply of making robots out of people, but of substituting robots for people and dispensing with the need for human labour altogether — all in the name of economic and technological progress. No wonder, then, that this second transition, like the first, is marked by social instability and economic crisis, "with the demand for work and workers equally diminished." Likewise, these conditions are again forcing "once honest and industrious" men and women into opposition in defence of society and their own humanity. In the process, moreover, people of the late twentieth century are beginning to shake loose from their inherited ideology of technological and economic progress that has for so long distanced them from their comrades of an earlier day.

In the mid-1970s, when the pressmen launched their predawn raid at the *Washington Post*, the action was perhaps inevitably isolated and futile. Fragmented unions each worked out their own formal agreements, co-operative programs, and research projects, and the best strategy still seemed to be one of deferred action and deference to authority. In the last few years, however, these conditions have changed dramatically, rendering agreements unenforceable, participation schemes mere collaboration in the administration of immiseration, and research projects less and less relevant. At the same time, these new conditions make other such predawn raids all the more necessary, promising, and likely.

The internationalization of the corporate economy, in manufacturing as well as finance, has given multinational capital unprecedented leverage vis-a-vis national labour movements and organizations. Thus collective bargaining has now become just a polite phrase for extortion, as labour has been compelled to yield. The recession of the early 1980s, with its rising unemployment, only magnified a more fundamental threat to the future of the labour movement. Capital's quest for greater control and profits, justified in the name of competitiveness and productivity and couched in the disarming rhetoric of technological progress, was now being facilitated through the now mature technology of automation. "If America hopes to match foreign competition," *Time* preached, "it may have to rely more heavily on automation." *Business Week*'s Harris Poll of late 1982 captured the true spirit of the age, however, noting that while executives were not interested in financing to rebuild inventory and did not intend to rehire laid-off workers, there was, nevertheless—in the midst of a recession—"heavy backing for capital investment in a variety of labour-saving technologies that are designed to fatten profits without necessarily adding to productive output." In 1983 *Fortune* simply heralded the "Race to the Automatic Factory," while *Time* substituted for its annual "Man of the Year" a more timely "Machine of the Year": the computer.

Confronted by this technology-based assault, battered by the economic recession, and confounded by its own (derived) commitment to technological progress, labour has been thrown on the defensive. In the process, unions have almost entirely abandoned the crucial struggles over technology and working conditions ignited by the rank-and-file rebellions of the late 1960s. Those workers who have continued to insist upon such shop-floor struggles have been dismissed by union officials intent upon maintaining dues-paying membership and keeping plants open, whatever the price in the present. Meanwhile, the "technology researchers" have abandoned even the pretence of dealing with the technologically based challenges of the present and have drifted ever more toward the development of technological alternatives for the future. However valuable all of this effort might some day prove to be, it is of little practical value to those now under immediate assault. Thus, in the face of an intensifying challenge, the capitulation of the unions, and the escapism of the experts, it is no wonder that workers in the shops have once again begun, increasingly, to take matters into their own hands. Having overcome the fatalism of technological determinism, they have now begun also to overcome the futurism of technological progress, and to shift attention back to the present. The resurgence of the rank and file, moreover, signals the return as well of direct action at the point of production. As Pierre Dubois predicted in 1975—the year of the

Washington Post raid and just as the economic crisis began to take hold—"All in all, we may say that unemployment is more likely to favor sabotage than not."

The *Washington Post* action was inescapably doomed, but not because the action itself was inherently wrongheaded or irrational, but because the conditions for its effectiveness had not yet materialized. The pressmen were fighting an awesome foe against overwhelming odds and were compelled by circumstance to do so alone. Since then, however, while changing conditions have in some ways certainly worsened labour's position, they have in other ways rendered such tactics more promising.

In the first place, the same diffusion of computer-based communications and control technologies and the same internationalization and conglomeratization of enterprise that have increased capital mobility and resourcefulness have also contributed to an increasing homogenization and integration of industry, with the result that workers throughout industry and throughout the world are now increasingly confronting the same problems. These new conditions, therefore, have created a basis for the recognition by workers (and, haltingly, by unions) of an identity of interest across industries and workplaces. Throughout the world of work—in factories, offices, design rooms, and warehouses, on the docks, in supermarkets and government bureaus, in aircraft, steel, auto, meatpacking, rubber, textiles, printing, and chemical plants—people face the same computer-based challenges. Thus, the conditions have emerged that make possible as never before liaison between crafts, technical and manual workers, factories and offices, and otherwise wholly different industries, and between workers in different nations.

Second, the same technology that has extended capital's reach and range of control has also rendered it more dependent upon highly complex, expensive, and notoriously unreliable systems and thus more vulnerable to worker resistance and especially to disruption through direct action. Increasingly, if belatedly, workers have begun to recognize the precarious position of management too during a period of rapid organizational and technical change.

Third and last, it is becoming increasingly apparent that this "window of vulnerability" of capital will not stay open forever. At some point the situation will become stabilized, the new systems will be sufficiently debugged, and the opportunities for opposition will be foreclosed. Moreover, in light of the current trend toward an ever-weaker labour movement, more people are beginning to understand that, however weak it might be now, labour is at present more powerful than it is likely to be in the future. Therefore, as one electrical worker at General Electric's Lynn, Massachusetts, plant put it, "You have to strike while the iron is hot." In short, the new conditions, while in some ways throwing labour on the defensive, have at the same time laid the basis for a rank-and-file resurgence across industries, with its accompanying emphasis upon the concrete situation, its orientation toward direct action at the point of production, and its preoccupation with the present. Already there are reminders of Ludd's warning: "Danger Looms."

> We will never lay down our arms till the House of Commons passes an Act to put down all machinery hurtful to the community— But we, we petition no more. That won't do—fighting must. Signed by the General of the Army of Redressers Ned Ludd, Clerk

In Norway, birthplace of technology agreements and a model for progressive unions throughout the world—where, as Leslie Schneider of Harvard Business School says, "More than any other country in the world unions, employers, and the state have tried to shape the direction of technological change at work"—workers have begun a search for new ways to deal with technology. In the face of the economic crisis, the unions have put the issue of technology on the back burner: "These matters of new technology are now in the background," one staffer for the Norwegian Chemical Workers Union noted. Kristen Nygaard, Norwegian Computer Centre scientist and pioneer of the pathbreaking Iron and Metal Workers Union technology agreements, observed also "a spreading sentiment of the irrelevance of the old participation strategy in the face of current needs." The ineffectiveness of such a strategy has also become more than apparent. At Kongsberg Vapensfabrik, for example, site of one of the earliest technology agreements, the local Iron and Metal Workers Union has demanded, "All changes in working conditions, past practices, job content, qualification, and skills due to the introduction and use of new machines be discussed and negotiated in advance." These demands have been rejected by management, and workers face the prospect of massive layoffs. At other sites, however, such ineffectiveness has given rise to new rank-and-file initiatives.

At Borregaard Industries (paper and pulp products) in Sarpsborg, for example, where formal agreements have proved similarly ineffective, a splinter group of ironworkers and metalworkers (unaffiliated with the national union) has begun "struggling toward a 'before the fact' approach to technological bargaining." According to Leslie Schneider, who in 1983 produced a revealing study of technology agreements in Norway, these workers "stopped" one system-design project twice "when they felt they had no power or real voice." In response to management plans to introduce a computer-based maintenance system that would have entailed a reduction in manning and tighter management control over those who remained, the workers countered with an alternative proposal that included group planning of work and job rotation of supervisors. Management rejected their proposal and the workers have decided to "block" the new system altogether until their alternative is accepted. "So far," Schneider reports, the iron and metalworkers "have relied heavily on the strength of the local union to block systems or project work that does not meet their demands." Meanwhile, in the city of Bergen, city government workers, frustrated by the ineffectiveness of technology agreements, have demanded and won a moratorium on the development of all new technological systems until management submits to them a long-term plan for technological change. The success of the moratorium was based upon an alliance between the city government union and the technical specialists who design the computer systems for the city, an alliance grounded upon a shared commitment to local control. The moratorium has given the unions and workers time to draw up their own "Policy on Technological Change," which has been critical for clarifying the union's ideas and goals internally and for refining their effort on the strategic level for their negotiations with management. Here, then, as at Borregaard, alternative plans for the future have been coupled with decisive action in the present. It is understood that just as it is not enough to engage in immediate action without longer-range objectives and visions, so it too is not sufficient to have the goals without the active means for attaining them.

In Denmark, where unions have also had extensive experience with technology agreements and technology researchers, workers have begun looking for more effective mechanisms. For some years now, workers throughout Denmark have been demanding a veto right in all agreements on new technology, but have thus far been opposed by the central unions. Existing agreements, therefore, contain no provisions that allow the workers to veto, or reject, any new technology. In the view of many workers, agreements without veto power have no "teeth" and serve merely to facilitate the introduction of new technology. "If you go out to the members," LO (Danish trade union federation) research director Henning Tjornehoj observed in December 1982, you discover that "workers want to fight for the veto since agreements without it are useless." The central unions have dismissed such demands as the work of left-wing agitators and have argued that the approach would stop technology, undermine competitiveness, and therefore result in the loss of more jobs. But, Tjornehoj insists, the demands for the veto are not restricted to leftists but come from workers, regardless of political views. Without the veto power, he agrees, "Workers are faced with the choice of being hanged or being shot." Although he himself oversaw the Danish trade union-sponsored action research projects (such as DUE and PUMA), Tjornehoj concedes that "it is unrealistic to be effective in shaping technology" and warns the unions that if they do not take the lead in formulating a more effective approach to the challenge of new technology, the workers will begin to "take their fate into their own hands."

This has already begun. In 1982, municipal workers in the city of Farum, near Copenhagen, went on strike to demand veto power over the new technology, in opposition not only to local management but also to the central union and the central government. They backed down only after the central union and the social-democratic government threatened to cut off funds to the city. In one place, however, workers have actually succeeded in obtaining the veto. Ironically, this victory has been achieved by clerical workers in the state-run Business School in Copenhagen. There the workers discovered to their surprise that the management knew as little about the new office automation technology as they did. Taking the initiative, the local union forced the management into an agreement that permits the union to reject any new systems after a three-month trial period. So far, the union has succeeded in preventing the permanent introduction of any new equipment, on the grounds that the new technology deskills and displaces members of the workforce. Thus, while the professors at the Business School daily spread the gospel of salvation through automation, the workers at the Business School have taken the lead (in Denmark) by creatively responding to this threat and so far have prevented these missionaries from practising what they preach.

"England's loss was our gain," John Baker, former general secretary of one of Australia's postal-telecom unions, has observed, referring to the fact that many of the convicted Luddites were "transported" to Australia and thereafter had a considerable influence upon the development of Australian trade unionism. "Where the worker responses were active, positive, and assertive on their own immediate interests," Baker noted, "these attitudes flowed through the rest of society with rather positive consequences for most institutions of society." (Australia led Europe and the United States in mass unionism, the eight-hour day, social security, and social democracy.) This heritage is still alive in Australia. In 1954, for example, the postal-telecom unions refused a continent-wide automated telegraph system until the union had a chance to scrutinize it in terms of efficiency, social necessity, and consequences for those in the industry. In 1977 the Australian Labour Party called for a moratorium on uranium mining and treatment in Australia (following a nationwide strike a decade earlier by railway and transport workers over that issue). That same year saw another series of strikes by the postal-telecom unions over the introduction of a new system. "We won't permit the introduction of an electronic telecommunications network," they declared. "Our members will not move over for a computer." In their refusal to accept the new system, the union employed various forms of industrial action, including one that captured the popular imagination: "During the dispute, which the technicians conducted from inside the telecom systems, they cut-over the local-call system to the nation-wide long distance system and enabled subscribers to make unlimited long-distance calls for the price of a local call." Finally, in 1979, the Australian Council of Trade Unions voted to request the International Confederation of Free Trade Unions and the International Labor Organization to invite labour unions of all affected countries to "consider placing a five-year moratorium on all technological change." Baker recommended, "A little bit of creative Luddism might not be amiss until we sort things out." He observed:

> The developing consciousness of the Australian trade unionist illustrates the old challenge of the Luddites to the factory-owners: "you haven't any right to take over my tools and skills and build them into a machine [that] you, alone, own and whose products you, alone, sell in the marketplace." This old objection is being resurrected again as owners of technology and capital build the skills, experience, and knowledge of millions of office and factory workers into the micro-machine processes that make them unemployed.
>
> Like John Brown's Body, that spectre, that special understanding

of the Luddite Martyrs marches on, coming back to haunt the heirs of those who transported them in irons to the shores of Botany Bay, coming back to haunt Westminster until, perhaps, some Labour MP dares to rise, as once did Lord Byron in the House of Lords, to honour the Luddite Martyrs in the way their consciousness and sacrifices still warrants.

In 1982 strikers in Australia began distributing stickers that read: "SMASH THIS MACHINE."

In England, meanwhile, site of the original Luddite uprising, design engineer Mike Cooley, one of the most active members of the Lucas Aerospace Workers Combine Shop Stewards Committee until he was sacked in 1980s, has seconded John Baker's appeal for "a little creative Luddism." As Cooley noted, "The real tragedy is that with the frantic drive forward of the new technology, we lack the time to examine the cultural, political, and social implications before infrastructures are established which will effectively preclude any examination of alternatives." Cooley has also welcomed a moratorium on new technology as being consistent with the efforts of the Lucas workers' strategy. This might come as something of a surprise to many of those in the United States who have been inspired by the creative initiatives of the Lucas workers, in particular their alternative corporate plan, their development of alternative products and processes, and their programs for civilian conversion of military industry. The emphasis, it might appear, has been upon developing alternatives to corporate technological development, not rejection of new technologies. But observers in this country and elsewhere have concentrated only on one-half of the Lucas story and ignored the other. Throughout the development of the Combine Committee, the formulation of the alternative plan, and the endless campaigning, there has been consistent and determined industrial action at the point of production. All along, the more visible parts of the Lucas strategy were rooted in and undergirded by strikes, occupations, slowdowns, and other forms of industrial action for which British workers are renowned and respected.

One of the most significant achievements of the Combine Committee, for example, barely known outside Lucas, was a successful moratorium on the introduction of new technology. In the only reference to it in all of the extensive Lucas-related literature, Hillary Wainwright and Dave Elliott have described how in 1980-81 the Combine Committee "succeeded in coordinating a moratorium on the introduction of new machinery at a time when Lucas Aerospace management was hoping to move rapidly into computer-aided design and computer-aided manufacturing systems at major sites." The moratorium had two objectives. First, it was a way of getting management to negotiate centrally on new technology with representatives from all sites and all unions rather than with particular unions at each individual site as management intended. The Combine Committee understood that a piecemeal introduction of the machinery would weaken the ability of the unions to develop a truly effective strategy for controlling the pace and the terms of the new technology's introduction. The Combine Committee reflected the interests of both production and technical workers at all sites, and the moratorium was a practical means of maintaining this unprecedented solidarity and strength. Second, the Combine Committee intended to use the time made possible by the moratorium to involve all of the shop stewards committees in the formulation of policies that would be the basis for central negotiations. According to Wainwright and Elliott:

> The moratorium lasted for nearly a year in some sites and longer in others (at Burnley it held for eighteen months). During this period the Combine Committee and CAITS (the Centre for Alternative Industrial Systems set up by the Lucas workers at the North London Polytechnic) carried out an extensive investigation and debate on the consequences of new technology for workers in Lucas Aerospace and the policies through which the benefits of new technology could be controlled by those who manufacture it, use it, and consume its products, rather than by those who own it. The moratorium did not hold consistently enough to force management to conduct central negotiations, but it did lead to several good site-level agreements.

After the moratorium some Lucas workers devised other forms of industrial action that they found more effective than the refusal to allow the introduction of new equipment. At Birmingham, for example, workers realized that the forcible rejection of new technology at their site, in the absence of total rejection at all sites, simply meant that the company would place the equipment elsewhere, at their expense. Therefore, they have adopted another approach. They allow the company to bring in the new equipment, install it in concrete on the factory floor, test it, and work out the "bugs." Then they take it over, to prevent anyone from using it and to prevent the company from removing it. As long as the equipment is outside, the workers reason, the company can control it, whereas once it is in the factory, the workers can control it. The objective of this strategy is to reduce the flexibility and mobility of corporate management and, at the same time, to idle expensive equipment and thus force plant management to negotiate reasonably.

The effectiveness of the strategy depends upon a full understanding of the company program for introducing new equipment. The workers must determine the "point of no return" for the company, at which time management must either move forward with new equipment or sacrifice the cost-effectiveness of existing equipment. Most systems are introduced in successive phases of system integration, where the efficiency of one phase is ultimately dependent upon the completion of the following phase. Given the enormous expense of new computer-based manufacturing equipment, interruption of this program at critical points can prove extremely costly to the company and thus helpful to workers in their negotiations. (In addition to this overt strategy, the workers here as elsewhere routinely feed irrelevant information into the company's central computer and otherwise forestall cost-effective utilization of the expensive equipment, alert to the fact that automation renders management more dependent upon the workforce, not less, and therefore more vulnerable to worker resistance.)

Of course in England too workers are daily confronted with management threats to close or move plants in the event of opposition to company modernization plans. But the workers at Birmingham and elsewhere figure that they lose either way—through rationalization and automation or plant shutdowns and capital flight—if they permit the company to proceed unhindered. Thus they have resolved to fight whenever and wherever they can—on their own turf, in their own terms, and now. The major problem with formal procedures and agreements, electrician John Routley, Combine Committee representative from the Lucas Birmingham site, explained, is that they disorient workers by forcing them to abandon reliance upon their own instincts for battle, instincts developed over time from experience with industrial action. "We see the Combine as it stands as a 'think tank' which draws on its members at all levels to work out . . . strategy," Routley told Wainwright and Elliott. "But what it does not have is the ability to carry out that strategy to its full potential. And that problem boils down to the fundamental problem of industrial strength." Without this industrial strength, he concluded, "All you end up with is 'research.'"

In Germany too, workers have begun to understand that there is a war on, and that strategies of co-determination, participation, and alternative technology are grossly inadequate in the wake of the current corporate restructuring, economic crisis, and technological assault. "The dockworkers know that the new technology will never benefit them," one dockworker from Hamburg insisted in 1982 at a historic meeting of rank-and-file dockworkers from throughout Europe. Likewise, at a meeting in Hamburg a group of workers from several industries in Northern Germany tried to explain themselves to some sympathetic computer scientists. They said they had "different expectations" than the scientists. "We are concerned less about some far-off future than the here and now." Inevitably, workers are expressing their dissatisfaction with outdated strategies. "Co-determination in personnel planning only made sense in times of expansion, not in times of redundancy," one works council member from a Hamburg printing firm argued. "Workers should not take part in redundancies." A colleague of hers from a Hamburg publishing company agreed that "Formal agreements don't actually enforce controls on management," because employers routinely violate the agreements anyway: "In practice we are helpless." Thus, as workers have begun to get involved in struggles with the companies directly rather than through formal works councils, as another printing worker pointed out, "There is now a shift toward resistance to technology."

One major focus of such resistance to technology in Germany has been so-called personnel information systems (PIS), computer-based management systems that enable managers to compile a mass of work-related and personal information about each employee. In 1979, workers at Daimler Benz in Stuttgart declared their opposition to PIS, organized their resistance, and forced the company into an agreement about how it would be used; the Daimler Benz workers failed to prevent the installation of the system, however, to their subsequent regret. But this resistance sparked a similar movement against such systems at Opel and elsewhere and led to a remarkable series of meetings in Frankfurt and Hamburg in 1981. Among those in attendance were rank-and-file workers from such industries as auto, aircraft, banks, insurance, printing, publishing, oil refining, and the docks, as well as municipal workers. The theme of the Hamburg meeting (which I attended) was *"Nein zu PIS,"* with the emphasis definitely on the Nein rather than the PIS. Focusing upon personnel information systems was tactical; the organizers hoped that the widespread abhorrence of these systems could be extended to other equally but less obviously deleterious technologies. The discussions were marked by a general scepticism about formal arrangements over technology, a sentiment reinforced by the announcement that workers at Opel had obtained a company document containing lists of information that had been prohibited by the existing agreement. "Once we enter into discussions about changing the introduction of the technology," one aircraft worker commented, "we cannot say *Nein*." Yet *Nein* had become the order of the day, out of necessity. As one dockworker put it, "Now is the time to act; if we don't act now we will only lose power."

The resistance to technology from below has forced the union officialdom to adopt an unprecedented stance of opposition to PIS. In 1980 the Public Services Union demanded that PIS systems be prohibited; in 1981, H.O. Vetter, former DGB (central labour confederation) president, acknowledged, "We must not admit everything that is technologically possible." Finally, in 1982, the Federal Congress of the DGB in Berlin, in a dramatic turnaround for this progressivist union, issued Resolution Number 7 demanding that PIS systems be forbidden by the state. But workers throughout Germany understand all too well that such union declarations, while important and indeed historic, will not in themselves suffice. (Workers also recognize that the unions themselves have introduced such systems in union-owned companies and that the unions have begun to use such systems to keep tabs on the activities of members.) Thus, the rank-and-file has begun to invent its own strategies. At the printing firm Bauer in Hamburg, for example, workers have refused to distribute or sign information cards required by the company to build up its PIS data base. (They realize, however, that such refusal is itself data that will find its way into the management machinery.) On the docks in Hamburg, workers have filed a class action suit against the company to try to get an injunction on the installation of a PIS. While realizing that this is only the beginning, the organizers of this action are using it to raise consciousness about the need to resist the technology (eight hundred workers signed the suit) and to question both the liberal and Marxist myths of salvation through technological "progress."

One manifestation of this rank-and-file resurgence in Germany, and of the growing resistance to technology in particular, is the increasing public mention of sabotage. An issue of a political journal for scientists and engineers was devoted entirely to a discussion of sabotage, and sales exceeded all expectations. At the Hamburg conference on PIS, those in attendance wore large, colourful buttons depicting people with wire-cutters cutting cables and people physically attacking and destroying computers and robots. Perhaps the most dramatic discussion of sabotage was sparked by the case of Ulrich Briefs. A computer scientist who served for many years as an adviser on technical matters to the metalworkers union (I.G. Metall) and, more recently, the DGB, Briefs has also been unusually close to rank-and-file groups throughout Germany, travelling widely to engage in plant-level discussions and to help devise strategies for dealing with the new technologies.

In December 1981, as companies in Germany accelerated the pace of technological change, Briefs gave a speech publicly criticizing the DGB for having done too little too late. Speaking to a local union audience, Briefs noted, however, that all was not yet lost. He pointed out that computer systems do not make the company invincible; indeed, they intensify management's need for access to and control over data and they increase management's dependence upon the reliability of hardware and co-operation of the workforce. He suggested that strategies might include two forms of sabotage: "soft" and "hard." By "soft sabotage" Briefs meant anything that obstructs and distorts the information-processing system. He urged that workers should try to clog the system with extraneous information (as the Lucas workers were already doing) or overload it by making excessive simultaneous demands on it (such as all of them complaining about their paycheques at once). He emphasized what he considered to be the gravest danger of computer systems, the trend toward total integration, and recommended that workers should insist upon interrupting such integration in any way possible (such as demanding that non-automated processes be inserted between automated ones). As for "hard sabotage," Briefs explained in an understated way that computers do not like tea, coffee, Coke, or iron powder.

Almost immediately after his speech, the DGB fired Briefs. Although the action was presumably prompted by Briefs's public criticism of the trade union confederation, the charge against him was "Luddism," the advocacy of allegedly "anticonstitutional" criminal acts judged in violation of the principles upon which trade unionism was grounded. In response to these charges Briefs argued that, first, capitalists are the real saboteurs because they routinely destroy billions of dollars in equipment when they "retool" or close plants, and, second, that sabotage can be a very human act if it is directed against such antihuman devices as neutron bombs and other military hardware. More important than Briefs's arguments in his own defence, however, was the widespread support for him among trade union members. After only two months, during which time tremendous pressure was placed upon the DGB by local and district level union delegates (especially from the metalworkers), Briefs was reinstated. Some were expressing support for free speech and democracy in general rather than any particular sympathy with his remarks. Many, however, either agreed with Briefs's approach or recognized that it was time to broaden the discussion and seriously entertain the possibility of new and bolder strategies.

Over the Easter weekend in 1980, in the French city of Toulouse, there were unprecedented raids on the computer centres of both Philips Data Systems Corporation and Honeywell-Bull. The damage was extensive and *The New York Times* reported that "officials said the damage was done by experts." The *Times* quoted a police inspector: "They knew exactly how to erase programs from tapes, how to destroy the electronic filing systems." *Newsweek* reported that one technician (reflecting the peculiar logic of our times) exclaimed, "When they attack men, that I can understand, but machines — No!" The group taking responsibility for the raids, which called itself the Committee for the Liquidation and Misappropriation of Computers, explained itself in a letter to the Paris daily *Libération*: "We are computer workers and therefore well placed to know the present and future dangers of computer systems. Computers are the favorite instrument of the powerful. They are used to classify, control, and to repress. We do not want to be shut up in the ghettos of programs and organizational patterns."

"As proof of their involvement," the *Times* reported, "the group described the contents of (one) manager's desk, including a copy of a Rolls-Royce catalog." Meanwhile, in the Netherlands, according to a report in a Detroit newspaper, a professor of industrial robotics has concluded a study in which he found that sabotage of robots has become widespread and has assumed several ingenious forms. Workers routinely slow down the machines by feeding them parts in the wrong order, "repairing" the machines incorrectly, mislaying essential spare parts, or putting sand into the robot's lubricating oil. In one metal construction plant, the professor reported, production was reduced for more than six months because of worker resistance to the use of robots.

Finally, in the United States, people are only belatedly coming to realize that their society has become just another reservoir of factors of production for multinational capital. Resistance to the technological assault has only begun to surface in unorganized, fragmented, and covert ways. But there are signs that the ideological fiction of "labour resistance to change" is now becoming a defiant reality as workers learn the hard way that formal agreements about technology (such as those of the Auto and Communications Workers and those proposed by the Machinists) are barely worth the paper they are written on. In 1979, for example, in opposition to their own International, UAW local 60 autoworkers at the Ford River Rouge plant in Detroit declared that it was time to make "new technology a strikeable issue." In Lynn, Massachusetts, members of International Union of Electricians' Local 20 have begun to join together with their counterparts in locals of their own and other unions (such as the IUE and the IAM) from Schenectady, New York, Evendale, Ohio, Erie, Pennsylvania, and elsewhere to collectively confront the technological assault under way in one of the country's most automation-obsessed companies, General Electric. Whether or not this is an incipient Combine Committee, along the lines pioneered at Lucas Aerospace, remains to be seen, but there is little doubt that, whatever ultimate form the resistance takes, the battle lines are being drawn.

"I sweep up robot doo-doo," was the way one skilled machinist at G.E.'s Erie locomotive works described his recently deskilled job at a conference on the new technology in Lynn. The same technology that was being used to degrade his working life, he explained, was also enabling the company to move jobs elsewhere or eliminate them completely. "When they closed the division," he told his fellow G.E. workers, we realized too late that "we should have acted earlier and destroyed it ourselves."

"Technology does not benefit the workers one bit," a displaced Baltimore steelworker insisted, pointing out how technology was being used to degrade and destroy jobs and produce things that most workers, as consumers, will never be able to buy. He went on to describe in some detail how workers in his plant were turning to more direct ways of protecting themselves against the introduction of new technology.

With the new computer technology, "You can undermine some resistance but they can still beat you," a manager at a large United Technologies plant (Hamilton Standard) in Hartford, Connecticut, conceded. "They'll always find a way to beat you. They're smart!" As every manager in the United States knows, but few are willing to tell, sabotage is on the rise. "The workers out there don't have the luxury of fantasizing about the future," the same manager reminded two academic researchers. "They don't think like you guys do. They don't see any inexorable technology. Survival is the thing. They think immediate."

One of the more innovative, and symbolic, acts of displeasure with new technology surfaced in reports of an incident in the U.S. Justice Department. In May 1983 a manager noticed that a word processor was not functioning properly. Upon closer inspection, he discovered that the screen and keyboard of the machine were saturated with urine. (Apparently, this readily available substance has the same effect upon computer equipment as tea, coffee, Coke, and iron powder.) With characteristic paranoia, the Justice Department management collected a sample of the offending substance and dispatched it at once to the Center for Disease Control in Atlanta, presumably in an effort to track down the resourceful operator. All they were able to learn, however, was that the source was female and free of social diseases. Meanwhile, that same month, a Detroit newspaper responded to *(Time* magazine's "Machine of the Year" (the computer) with an announcement of its own "Tool of the Year": the sledgehammer. Elsewhere, a new (old) phrase was coined to capture the spirit of the times: "SMASH MACHINES, NOT PEOPLE."

Has Ned Ludd returned? While it is not yet possible to assess the full significance of this mounting worker resistance to so-called progress, there certainly are signs of it everywhere. Likewise, there is abundant evidence of latent popular support for such resistance. The popularity of movies such as *Return of the Jedi* or *War Games*, for example, stems less from their celebration of technological prowess than from their mockery of technological extravagance, hubris, and vulnerability. Audiences are most animated when the kid in *War Games* uses his rudimentary knowledge about electronics and computers to outwit Pentagon technocrats and, of course, the phone company. Similarly, audiences respond to *Return of the Jedi* with the greatest delight when the supersophisticated Death Star warriors are undone, not by the equally sophisticated (and vacuous) heroes, but by the sticks and stones — and laughter — of the "primitive" (and most human) Ewoks. The producers of these films exploit not only the widespread resentment and anxiety about high technology (and the alienated lives and horrible dangers that accompany it) but also the genuine pleasure, the recovered sense of dignity, and the surge of power (albeit vicarious) when all the fancy gadgetry of those in command is put in its proper, diminished, place.

How one might finally interpret all of this is less important than how one ought to respond to it in the present, to seize and enlarge upon the opportunities it creates. It would be presumptuous and, indeed, contradictory to the main thrust of this book to try to suggest a full-blown program for labour. As I have emphasized from the outset, much of our confusion and paralysis in the face of the current corporate-sponsored technological assault has resulted from just such a removal of the technology question from the point of production, a substitution of futuristic and programmatic vision for workers' present-tense tactics and strategies. If any recommendations might be made, then, they must of necessity be restricted to the typical readers of a book such as this: "intellectuals," those with the luxury of reflection and access to the media. What might these people do on the ideological battlefield that would complement and enhance rather than supplant and stifle worker resistance to an alien and devastating progress?

In essence, if workers have begun to smash the physical machinery of domination, so responsible intellectuals must begin to deliberately smash the mental machinery of domination. They must strive to overcome — in themselves as well as in others — the collective fear of being human and free, a fear now reified and ratified in fixed ideas and solid-state circuitry. To do this, they must champion a new common sense that insists without compromise upon the primacy of people's lives over the strange and estranging myths of automatic destiny. The intellectual task is one of recovery, reclamation, and reminders: of who and what we are and of what is being lost. If people are to be encouraged in what they already partially see (as evidenced by their joyful identification with the Ewoks and the *War Games* hero), intellectuals must affirm outright, without qualification or hesitation: progress is a lie. Only then will more people be able to think, say, and act upon what they already know, without fear of isolation, ridicule, or repression. Responsible intellectuals, in short, must struggle in their own realm to gain legitimacy for worker resistance to progress. They must change the terms of debate and extend the range of respectable discourse and behaviour in order to ensure that those who choose to resist need never act alone.

It is an awesome challenge. When confronted with an identical task at an earlier time, Lord Byron too had second thoughts. He worried about his own reputation and dreaded that he might appear to his friends as a "sentimentalist" or, worse, as "half a frame-breaker" himself. "However we may rejoice in any improvement in the arts which may be beneficial to mankind," he wrote in a letter to Lord Holland shortly before his maiden speech, "we must not allow mankind to be sacrificed to improvements in mechanism" especially when these result merely in "the enrichment of a few monopolists." But the young Byron toned down his speech in the House of Lords, yielding to parliamentary restraint and gentility. His finely crafted, moderated oration was well received by his peers in that polite assembly, but the frame bill passed overwhelmingly in spite of it. (See Appendix V.)

Only then did the poet turn to his ultimate weapon, verse, to champion the Luddites, and humanity's cause. His defiant challenge to the masters of markets and machinery appeared anonymously in the next day's *Morning Chronicle*, as the bitterly ironic "Ode to the Framers of the Frame Bill."

> Oh well done, Lord Eldon! and better done Ryder!
> Britannia must prosper with counsels like yours;
> Hawkesby, Harrowby, help you to guide her,
> Whose remedy only must kill ere it cues.
> Those villains the Weavers, are all grown refractory,
> Asking some succour for Charity's sake—
> So hang them in clusters round each Manufactory,
> That will at once put an end to mistake.
> The rascals, perhaps, may betake them to robbing,
> The dogs to be sure have got nothing to eat—
> So if we can hang them for breaking a bobbin,
> 'Twill save all the Government's money and meat.
> Men are more easily made than machinery—
> Stockings fetch better prices than lives—
> Gibbets on Sherwood will heighten the scenery,
> Showing how Commerce, how Liberty thrives.

As we approach the calamity of the second industrial revolution, intellectuals have again to take up the vital struggle begun by Byron amidst the calamity of the first. As witness to the birth of industrial capitalism, Byron defied the nascent laissez-faire ideology that perverted liberty and invention and turned them against society. Today, we witness the final maturation of a still antisocial capitalist system in which liberty and invention have hardened into a monopoly of power sanctioned by shibboleths of automatic progress. A century and a half of obeisance to progress has dimmed our memory, impoverished our imagination, and dulled our sense of outrage and violation. It is thus more difficult than ever (and more urgent) to regain the courage and composure to speak candidly about what is happening, and about what must be done to stop it.

At least five tasks await the committed intellectual: to shift the burden of proof; to create the space to say no; to develop the means of resistance; to invent an alternative future that is moored in the present; and to transcend the myth of the machine, the fetish for technological transcendence, in order to make way for political opposition.

The distinguishing characteristic of hegemonic ideologies is that they require no proof or argument; their validity is assumed, understood, and ratified by convention, norms, and accepted bounds of respectable discourse. Thus, those who challenge this dominant set of ideas are typically the ones who must bear the burden of proof, a burden that, in this setting, actually defies argumentation and evidence. What is required, therefore, is somehow to shift the burden of proof to those who subscribe to, promote, and profit from, this ideology. If they can be forced to prove the validity of their ideas, their very attempt would be doubly defeating: first, because assuming the burden of proof would undermine the automatic acceptance of their position and, second, because, not having had to prove themselves before, they would invariably demonstrate the weakness of their reasoning and the fact that, given the available evidence, their assumptions cannot compellingly be defended. Ideologies are habits of thought that defy thought and enable people to avoid thought. Forcing that burden upon them (and upon ourselves) brings blindly accepted assumptions to consciousness, and breaks the habit.

It is not necessary to demonstrate that accepted assumptions are wrong, but merely that they are ambiguous. Once the ambiguity has been established, further evidence or argument becomes necessary for both sides. Intellectuals need to raise questions about technological development: about its design, its deployment, the reasons for its introduction, its technical and economic viability, and the causal connections between investment, innovation, productivity, competitiveness, and social welfare. Any serious present-tense assessment of new technologies would readily reveal the fragility of assumed justifications. (See Part Two.) Contrast this with the clear-cut social costs entailed in the corporate-sponsored application of new technology, including structural unemployment, social dislocation, job degradation, worker deskilling, and political instability. It must fall to the ideologues of progress to prove, rather than simply assume, the benefits *before* they are allowed to proceed.

The first industrial revolution was recognized as such only in retrospect (the term was not coined until the transformation itself had been completed). The second industrial revolution, in contrast, has already been identified in anticipation of the event. Thus, we have a luxury, denied our predecessors, of entering the transition with our eyes open and with the advantage of a precedent. Whatever blindness remains is self-induced.

Among our inherited blinders is the identification of simple technological advance with social progress, an idea espoused by liberals and socialists alike. Late-twentieth-century North Americans need not be reminded that this belief is suspect and invites a fundamental re-evaluation. Given the all-too-important social costs, military, ecological, and socioeconomic, it would be wise to call a halt to rapid, undirected technological advance, if only temporarily until we regain our bearings. But we are confronted immediately with another inherited blind spot, the conviction that technological advance cannot be stopped because "You can't stop progress." In reality, this is a bizarre and relatively recent Western notion, invented to disarm critics of capitalism, and is readily refuted by reference to centuries of socially interrupted technological development. "Protective" regulations of one sort or another have long served to buffer society from disruptive changes; the Luddites themselves appealed to this time-honoured tradition, which assumed the supremacy of society over mere economic activity and technological contrivance. This consistent characteristic of human society was eradicated only within the last few centuries by the rapacious champions of laissez-faire, who succeeded in putting "things" in the saddle, to ride mankind. It is necessary now to remind ourselves of this lost tradition and confidently to reassert it. To the dictum, you can't stop progress, we must learn to respond: of course you can.

There are those who have tried to undo our modern fatalism, with little success. Norbert Wiener, the father of cybernetics, appealed in the 1940s and 1950s for some slowing down of the pace of automation and warned against an impending catastrophe for labour (he erred in his prediction, but only in terms of time: he was off by a decade or two). John Parsons, inventor of numerical control and the acknowledged (by the Society of Manufacturing Engineers) "father of the second industrial revolution," called also for a "moratorium on technological advance, to provide time for reflection and a search for socially responsible ways to proceed." Both protests were dismissed without a hearing. It is now urgent to revive their efforts and extend them.

One possible strategy might be to illustrate that, despite our espoused deference to technological advance and economic ends, our society routinely accepts certain limits on both. For example, we are learning to live with environmental constraints on both economic and technological activities. We would also now reject the reinstitutionalization of slavery even if it could be shown to enhance our international competitiveness. Yet if undue environmental degradation and the institution of slavery are today unacceptable, the capital flight and technological displacement that cause social dislocation are not. How might they become so? We have environmental impact statements, why do we not have employee impact statements? Required by all employers who wish to introduce new methods, these would demand an assessment of *and solutions to* likely problems before implementation. "We protect the fish," one electrical worker at G.E. Lynn observed, "why not the people?"

In April 1983 the United States Supreme Court ruled that states have the right to "allow the development of nuclear power to be slowed down or even stopped for economic reasons" and, in a minority opinion, two Justices went a step further and argued, "A ban on construction of nuclear power plants would be valid even if its authors were motivated by fear of a core meltdown or other nuclear catastrophe" (for example, nuclear waste hazards). Might this decision serve as a precedent for banning the use of robots pending the solution of the problem of unemployment? In recent years the California Rural Legal Assistance (CRLA) has sued the University of California, on behalf of farmworkers and small growers, in an attempt to prevent further publicly funded development of agricultural mechanization that serves only agribusiness at the expense of those workers and smaller farmers. The suit has been derided as an attempt to halt progress, and the CRLA lawyers have been denounced as Luddites. Ways must now be found to follow their lead.

Saying no to such technological innovation serves two purposes at once. First, the call to stop progress reminds us that we have been caught up in something moving, something we never started or ever decided to participate in. On the intellectual level, then, saying no brings our collective compulsion to consciousness and permits us to begin to proceed on a more rational basis, with our eyes open. Second, saying no does not so much arrest human history as call into question the current form of development and change the rules of the game in the present.

In line with the smashing of mental machinery, intellectuals must strive to overcome their own and others' undue reverence for, and deference to, physical machinery, in order to develop criteria, defences, and devices for effective resistance to technological change. No one is against "technology," despite the frequently heard charge, because technology as such does not exist. Technology exists only in the particular, as particular pieces of equipment in particular settings. Therefore, if opposition to technological progress is to be meaningful, if lost options are to be reassessed in the light of new conditions, criteria must be developed for selecting which technologies ought to be stopped. Technologies might be opposed, for example: if they degrade people and diminish their freedom and control without any apparent economic or other compensating benefit; if their technical and economic viability is ambiguous but they pose serious social problems; or if they are clearly viable in the narrow technical or economic sense but are nevertheless destructive for society as a whole. Similarly, a technology might be elected for opposition if management depends on it heavily. Such opposition to technologies must be defended in the public sphere, and intellectuals might helpfully formulate persuasive defences. These might include a moratorium in order to buy time in which to develop socially responsible procedures for their introduction, the protection of existing organizations, the guarantee of livelihoods, or the preservation of communities.

Reclaiming the present must not necessarily entail an abandonment of the future. It remains an essential task to develop alternative social and political visions, rediscover roads not taken, and recast science and technology according to life-enhancing criteria. This has always been and remains a central challenge for intellectuals. But care must be taken to ensure that such projections never *substitute* for present strategies, but serve rather to complement, inspire, and perhaps guide them. The danger is not utopianism — we still need utopia — but the confusion of the future with the present. If we cannot afford to abandon the future in our preoccupation with the immediate, neither can we afford any longer to concentrate upon the future and surrender the present. The two must be joined.

One illustration of how this might be done has been offered by Mike Cooley in England. He suggests that the development of "human-centred systems" for production, such as those projects now under way at the University of Manchester, must be coupled with shop-floor organizing and struggle. According to Cooley, the new technological systems are meaningful politically only if workers can be inspired to demand that they be substituted in the present for management-designed systems, and inspired to strike or take other direct action in support of this demand. Without such action, research projects are doomed to academic irrelevance.

If opposition to technological progress helps us overcome our infantile dreams of technological salvation, it enables us also, finally, to transcend the technological mystification of power in our society. For technology has never really been the problem, nor will it ever be the solution. Technology does not by itself destroy democracy, nor does it bring democracy into being. If we have become a politically regressive society, as Sheldon S. Wolin suggested in the first issue of the journal *democracy*, it is not because of the politics of technology but because, "hypnotized" by ideologies of progress, we have substituted technology for politics. The ultimate challenge posed by the current technological assault, therefore, is for us to learn not only to put technology in perspective but also to put it aside, to make way for politics. The goal must be not a human-centred technology, but a human-centred society.

This demands, as it always has, a confrontation with power and domination. If the lessons of the Luddites are instructive in this regard, so too is the observation of that English radical who, in 1835, summed up the matter parsimoniously, and in a manner still appropriate today:

> The real grievance is neither more nor less than the subjection of the labouring to the monied classes, in consequence of the latter having usurped the exclusive making of the laws. Rents, tithes, taxes, tolls, but above all profits. Here is our distress explained in five words, or to comprise all in one, it lies in the word Robbery.... Machines indeed.

(Chapters 1, 2, and 3 of this book are slightly revised versions of a three-part article that appeared in *democracy*, in the spring, summer, and fall 1983 issues.)

SOURCES

"America Rushes to High Tech for Growth, "*Business Week*, March 28, 1983. For the text of Byron's speech in the House of Lords, see Appendix V. On the *Washington Post* episode, *Time* quoted in Rosalind Williams, "The Machine Breakers," *Technology Illustrated*, July 1983; Ben A. Frankly, "Washington Post Is Shut by Pressroom Vandalism," *The New York Times*, October 2, 1975; Robert H. Kaiser, "The Strike at the Washington Post," *Washington Post*, February 26, 1976; "The Washington Post," Harvard Business School case study, Harvard Business School, 1976; interview with *Washington Post* labour relations director Lawrence Wallace, June 8, 1983. R.O. Ayres and S.M. Miller, *Robotics, Applications and Social Implications* (Cambridge, Mass.: Ballinger, 1981); Business Week/Harris Poll in *Business Week*, December 13, 1982; "Machine of the Year," *Time*, January 1, 1983; Gene Pylinsky, "The Race to the Automated Factory," *Fortune*, February 21, 1983; "The New Economy," *Time*, May 30, 1983; Pierre Dubois, *Sabotage in Industry* (Baltimore: Penguin, 1979); Minutes of New Technology Conference, IUE Local 20, Lynn, Massachusetts, June 1983; Ludd's warning in Leslie Kipling and Nick Hall, *On the Trail of the Luddites* (Hebden Bridge, England: Pennine Heritage Network, 1982); Leslie Schneider, "Technology Bargaining in Norway," unpublished paper prepared for the Ministry of Local Government and Labour, Oslo, Norway, March 1983; interview with Gunnar Myrvang, Norwegian Chemical Workers Union, October 1982; interview with shop steward in the Kongsberg Vapensfabrik, October 1982; Kristen Nygaard, "Prospectus for Unite Project," typescript, and discussion with Nygaard, October 1982; Henning Tjornehoj, speech, Technical University of Denmark, December 8, 1982; discussion with Lauge Rasmussen about events in Fanum, fall 1982; interview with Jane Hjotte Andersen, chair, Clerical Workers' Union Local, Copenhagen Business School; John S. Baker, "The Luddite Tradition in the Techno Challenge," typescript, 1979, courtesy of John S. Baker and Mike Cooley; Mike Cooley, "A Bit of Creative Luddism Would Not Be Amiss," *Computing*, February 10, 1983, p.14; Mike Cooley, speech, Technical University of Denmark, November 11, 1982; Hillary Wainwright and Dave Elliott,

The Lucas Plan (London: Allison and Busby, 1982, pp.214-15; interviews with Mike Cooley, Ernie Scarbrow (former secretary, Lucas Combine Shop Stewards' Committee), Hillary Wainwright, John Routley, and CAFRS director Mike George; *"Nein zu PIS,"* conference minutes, Hamburg, December 1982.

Interviews with dockworkers Gerd Muller and Jorg Wessels, Hamburg, September 1982; discussions with Heidrun Kaiser and colleagues, FORBIT, Hamburg, December 1983; interview with Ulrich Briefs, December 1982; "A War on Rench Computers," *Newsweek*, April 28, 1980; Frank J. Pfial, "Raiders in France Destroy Two Computer Centers," *The New York Times*, April 11, 1980; Study on robot sabotage in the Netherlands, cited in *Fifth Estate* (Detroit) 18, No.1 (Spring 1983); interview with systems manager, Hamilton Standard, Hartford, Conn., 1978; Byron quoted in Dora Redmond, *The Political Causes of Lord Byron* (New York: Russell and Russell, 1924); anonymous (Byron), "Ode to the Framers of the Frame Bill," *Morning Chronicle* (London), March 2, 1812, reprinted in Redmond, *Political Causes*; Norbert Wiener, letter to Walter Reuther, August 13, 1949, in Wiener papers, MIT archives; Norbert Wiener, *The Human Uses of Human Beings* (Boston: Houghton Mifflin, 1950); interviews with John Parsons and Cart Parsons, Travers City, Michigan, 1980; Frank Emspak, speech, New Technology Conference, Lynn, Mass., June 1983; *Pacific Gas and Electric Co. et al. v. State Energy Resource Conservation and Development Commission et al.*, U.S. Supreme Court decision 81-1945, decided April 10, 1983; California Rural Legal Assistance mechanization case described in Albert Meyerhoff, "Agribusiness on Campus," *The Nation*, February 16, 1980, and discussions with Meyerhoff and Ralph Abascal of the CRLA; Sheldon S. Wolin, editorial, *democracy* 1, no.1 (January 1981).

Final quote from Maxine Berg, *The Machinery Question and the Making of Political Economy* (Cambridge: Cambridge University Press, 1980), p.286. For a discussion of how intellectuals can raise questions about technological development, see my *Forces of Production: A Social History of Industrial Automation* (New York: Alfred A. Knopf, 1984).

PART 2
AUTOMATION MADNESS: OR, THE UNAUTOMATIC HISTORY OF AUTOMATION

> Strange business, this crusading spirit of the managers and engineers, the idea of designing and manufacturing and distributing being sort of a holy war; all that folklore was cooked up by public relations and advertising men hired by managers and engineers to make big business popular in the old days, which it certainly wasn't in the beginning. Now, the engineers and managers believe with all their hearts the glorious things their forbears hired people to say about them. Yesterday's snow job becomes today's sermon.
>
> — Kurt Vonnegut, *Player Piano* (1952)

Today's sermon is a print-out on the glories of computerized automation, of progress without people. And when the engineers and managers preach this gospel from electronic pulpits everywhere, they now truly believe their own snow job, and so too do we all. Together, we seek salvation in their scientific fantasies, calculate our chances with their egoistic economics, and accept their vision of the inevitable future as our destiny too. In the previous chapters I have tried to identify where these shared habits of thought came from. Here I want to examine this set of quasi-religious ideas more closely and critically in the hope of moving us beyond them. Because, whatever hardship and suffering we might experience as workers in the wake of automation, these ideas prevent us from responding in our own interests.

Our collective inheritance from the past, these thought processes are also reinforced by our actions in the present, by our deference to those in power and those "in the know," and also by our own behaviour as consumers, when we allow the delights and promises of new products to fuel our enchantment with and our fetish for technological salvation, and to numb us to its social costs. Today our survival demands that we take another look at our ideas about progress and confront the sobering realities that these ideas allow us to ignore. For only then can we begin to challenge the self-anointed apostles of automation whose social irresponsibility is matched only by their madness.

Put simply, we have come to believe in an automatic future, one driven and directed by an autonomous technological advance (technological progress) and leading inescapably to the best of all possible worlds (social progress). The first, we suppose, proceeds automatically and guarantees the second. Let us look at each in turn.

CHAPTER 4

AUTOMATIC TECHNOLOGICAL PROGRESS

As a result both of ignorance and incessant inculcation by our established institutions, we have all come to hold a rather simple, though deceptively straightforward, view of how technology develops. In essence this view is Darwinian; we believe that the process of technological development is very much like the biological evolution of the species through natural selection. Just as the Earth's creatures evolve according to the anonymous and automatic logic of survival of the fittest, whereby only those forms best adjusted to the rigours of nature survive, so too do the myriad technological possibilities generated by human imagination and ingenuity pass through a competitive and thoroughgoing process of elimination, which guarantees that only those best suited to human purposes survive — as it were, naturally and automatically.

Now, because this view is ideological, deeply ingrained as a habit of thought, we rarely if ever actually think about it. But when we do, our half-conscious ideas look something like this: we assume that our technologies pass through two successive filters or screens, which automatically weed out the unsatisfactory contributions and allow only the best to emerge. The first is a technical or scientific screen vaguely composed of the work of scientists and engineers who, with their dedication to rationality and efficiency, methodically subject all technological possibilities to careful and objective scrutiny and select only the best solution to any given problem. It remains something of a mystery exactly how this is accomplished, but we rest assured that it is. The second screen is an economic filter, composed of two equally vague mechanisms. Upon successfully completing the technical test, the selected technologies are subjected to the no-nonsense, cost-accounting, profit-maximizing evaluation of hard-headed, practical businessmen who seek only the most economically viable technologies from among those deemed technically superior. The "real world" savvy of the businessman, we assume, corrects for the excesses of the less practical scientists and engineers. Finally, since even businessmen and their managers can make mistakes in judgement, we rely ultimately upon the fail-safe test administered automatically by the anonymous operation of the self-regulating market, which allows only the most economically astute businessmen to survive the rigours of competition and, with them, only the best technologies.

Thus, when we see a technology in regular and widespread industrial use, we confidently assume that it represents the best history had to offer, since it survived the successive tests of this process of natural selection. In this way we routinely dignify the present array of technology as the highest expression to date of so-called technological progress and, as such, we accept it as inevitable, a fact of life — beyond the realm not only of politics but even of thought and discussion.

But if we take a more careful and critical look at this seemingly inevitable process of technical development, we recognize at once that it is not really automatic at all, but political — something people plan for and struggle over. That is, we see that it is not some abstractly rational enterprise with an internal logic all its own, but rather a human effort that reflects at every turn the relations of power in society. This is the case for both the technical or scientific "screen" and the economic "screen" alike, as we will see.

BEHIND THE TECHNICAL SCREEN

When we look past the veil of mystery that enshrouds the work of technical people, we find that their activities reflect their relation to power at every point. Their link with power gives them power — it entitles them to practise their trade in the first place, to learn, to explore, to invent; it emboldens their imagination; and it gives them the wherewithal to put their grand designs into practice. In short, it is the support of those in power (in our society, those with money or those with political, military, or legal authority) that affords technical people the luxury to dream, to dream expansively (yet within well-understood limits) and to make their dreams come true (by imposing them on others). Although most scientists and engineers would admit to their dependence upon those with power, few would concede that this relationship actually influences the way they think about things. They would insist, rather, that they are guided in their work by technical considerations above all else, and that this is what makes their calling rational and thus compelling. Moreover, judging from my own experience working with and teaching technical people, I know that few engineers are deliberately out to destroy jobs or unions or to harm people in any way. Although, of course, in practice they must satisfy the requirements of their bosses, their clients, or their customers, ultimately they aim only to do the best work for the good of society. Yet, consistently, again and again, they turn out solutions that are good for the people in power (management) but often disastrous for the rest of us (workers). Can this be explained?

For one thing, few technical people have any contact whatsoever with workers; in their education and their professional careers, they typically communicate only with management. Not surprisingly, they tend to view the world pretty much as management does, whether they know it or not. They are taught and usually believe that this is simply the most objective way of looking at things, but it is, in reality, the view from the top, the perspective of those with power. To illustrate, let me cite one example from my teaching experience in the MIT engineering school. All the students were graduate engineers, quite talented and well-meaning. One year they had a project to study the hazards involved in the transportation of Liquefied Natural Gas (LNG) by truck throughout New England. LNG is a highly volatile and extremely flammable substance; if it escaped from the tanks in an accident it would ignite immediately and cause tremendous damage. So the students set out to examine this problem in depth and they did a very thorough, indeed exhaustive, job. They studied all the technical aspects of the problem, the engineering of the containment, the practical problems of loading the trucks, the scientific problems of the diffusion of escaping gas. To do this they spoke with nearly everyone involved — the shippers, trucking companies, local, state, and federal regulatory officials — they would have contacted the suppliers in Algeria if they had found it necessary. Yet, at the study's end, they had totally ignored the people most directly involved — namely, the drivers of the trucks. These people were readily accessible; they belonged to two unions that had local offices in the city, with listed phone numbers. Yet the students neglected to contact them. Why? This was not intentional, but it was not really an oversight either. It was ideological. The engineers viewed the workers either as just parts of their trucks or as an alien species on another planet; the workers were not in the same world as the engineers, managers, officials. It would have taken a tremendous leap of imagination and, indeed, an act of courage, for them to have crossed over the class line.

Not surprisingly, the perceptions and insights of the workers were missing from the study report, which naturally evolved — without any instruction — into a management document. Engineering education is like this. Engineering students are encouraged at every turn to identify with, emulate, and serve those in power and either to ignore or to manipulate all others.

A second example illustrates where this training leads. For seven years I investigated the history of automated machine tools. Much of the pioneering design and development work of these tools took place at MIT, and I spent many months pouring over the vast collection of documents from the ten-year project. I discovered that the engineers involved in creating this self-professed revolution in metal-working manufacturing had been in constant contact with industrial managers and military officers, who had sponsored and monitored the project. Yet I found not a single piece of paper indicating that there had been contact with any of the many thousands of men and women who work as machinists in the metal-working industry — those most knowledgeable about metal-cutting and, again, those most directly affected by the technical changes under development. Again, and for the same ideological reasons, the engineering effort was essentially a management effort, and the resulting technology reflected this limited perspective — the worldview of those in power.

Clearly, this closed world of technical people influences how these people think about things; from the outset they consider only those solutions that are compatible with power. Again, this assumption of power in the minds of engineers is rarely conscious, nor need it be. Exactly how it works to keep them on track is subtle but powerful, for it relies upon their own desires for recognition and power. Suppose, for instance, I were giving this talk one day and announced to my audience that I had developed an ingenious new technical system that would enable the audience to produce some widget in half the time it takes conventionally, that it included the latest, state-of-the art components and had been fully tested. The only thing the audience had to do was follow my instructions, do exactly what I told them for as long as I said. That is, a central operating feature of the system's design was that it gave me complete control over everyone else's activities. And suppose I was quite enthusiastic about this system and got very excited trying to convince the audience — insisting upon my right to make all of the decisions. Pretty soon, they would think me some kind of nut and perhaps show me the exit. Yet such systems are designed and sold every day; if I were to take that exact same design to Lee Iacocca or Henry Ford, or any top manager in industry, chances are they would consider me a genius, buy the system, and hire me to implement it. What exactly is the difference between the two situations, such that with the same invention, in the first case I would be ridiculed, and in the second hailed as brilliant?

The difference lies in the relations of power. In the first instance, I do not have the power to get the audience to follow my instructions, so my design seems absurd. In the second case, however, the executive knows that he could compel his employees to do as I say, and so the same design is considered not only viable but a breakthrough. To push this example a little further, suppose that the audience, instead of dismissing me as a lunatic, succeeded in engaging me in serious debate about the system and that, after a while, we had together worked out a compromise design that was satisfactory in every way but gave everyone equal say-so, a democratic design, so to speak. Now if I took this improved (and technically more challenging) design to the executive, he would be the one to dismiss it as absurd — what, a system that gives workers the same decision-making power as the manager? Nonsense. What are you, some kind of radical? The point is this: the viability of a design is not simply a technical or even economic evaluation but rather a political one. A technology is deemed viable if it conforms to the existing relations of power.

Engineers are not stupid people; however naive they might be about some things, they learn quite early on that in our society, the authoritarian pattern predominates in all institutions and workplaces. (Workplaces are either run autocratically by the boss or governed by labour contracts that give managers exclusive control over production and technical decisions.) So when an engineer begins to design a top-down technical system, he reasonably assumes from the outset that the social power of management will be available to make his system functionable. Such authoritarian systems are also simpler to design than more democratic ones, since they entail fewer independent variables, and this also makes them more appealing to designers. Finally, authoritarian systems satisfy the engineer's own will to control and offer the engineer a powerful place in the scheme of things. Thus, for all these reasons, new technical systems are conceived from the outset as authoritarian ones, perfectly suited for today's world. With little forethought and no malice to speak of, engineers routinely draw up designs and construct systems that concretely reinforce the power of those they serve. In the process, their own interests, ambitions, and compulsions become intertwined with and indistinguishable from those of their patrons, and these shared fantasies of omnipotence shape what they do. Never are all possibilities entertained and soberly evaluated, as the Darwinian idea of technological progress suggests; only those that are compatible with the authoritarian position and disposition of those with the power to choose.

When I studied the history of industrial automation, all of this became very clear to me. I found that while technical and economic considerations were always important, they were rarely the decisive factors when it came to what was ultimately designed and deployed. Behind the technical and economic rhetoric of justification I consistently found other impulses: 1) a management obsession with control; 2) a military emphasis upon command and performance; and 3) enthusiasms and compulsions that blindly fostered the drive for automaticity.

CONTROL: AN OBVIOUS OBSESSION

Many academic studies today purport to describe and explain the advance of industrial automation, but few ever even mention a major impulse behind that advance: management's obsession with and struggle for control over workers. Any scholar who so much as suggests that such a motivation exists is typically derided as a simple-minded conspiracy theorist, and his or her work is dismissed without a hearing. The rest, wanting to appear sophisticated, construct elaborate theories in their effort to avoid the obvious, which is much like trying to describe the action in a boxing match while pretending there is only one fighter in the ring. Small wonder, then, that academic treatises on this subject have so little to say to experienced workers. For, as anyone who has ever worked for a boss understands too well, management is concerned with one thing above all else, and that is staying in control. However much this control might be justified in the name of economic efficiency, with the self-serving claim, belied by nearly every sociological study of work, that centralized management authority is the key to productivity, the truth of the matter is that control is less a means to other ends than an end in itself. Indeed, as my own experience as a worker taught me, and my detailed study of automation at a large General Electric plant demonstrated beyond any doubt,[1] managers will in the end do what is necessary for them to remain managers, whatever the technical, economic, or social costs. To this end, they consistently solicit and welcome technologies that promise to enhance their power and minimize challenge to it, by enabling them to discipline, deskill (in order to reduce worker power as well as pay) and displace potentially recalcitrant workers. Perhaps more than any other single factor, this explains the historical trend toward capital-intensive production methods and ever more automatic machinery, which have typically been designed with such purposes in mind.

This is an old story, really, one perhaps best told by Andrew Ure, an early apostle of industrial automation, in the 1830s, back at the dawn of the industrial revolution:

> In the factories for spinning coarse yarn ... the mule-spinners [skilled workers] have abused their powers beyond endurance, domineering in the most arrogant manner ... over their masters. High wages, instead of leading to thankfulness of temper and improvement of mind, have, in too many cases, cherished pride and supplied funds for supporting refractory spirits in strikes, wantonly inflicted upon one set of mill-owners after another.... During a disastrous turmoil of [this] kind ... several of the capitalists ... had recourse to the celebrated machinists ... of Manchester, requesting them to direct [their] inventive talents ... to the construction of a self-acting mule. Under assurance of the most liberal encouragement in the adoption of his inventions, Mr. Roberts ... suspended his professional pursuits as an engineer, and set his fertile services to construct a spinning automaton.... Thus, the Iron Man, as the operatives fitly call it, sprung out of the hands of our modern Prometheus at the bidding of Minerva—a creation destined to restore order among the industrious classes.... This invention confirms the great doctrine already propounded, that when capital enlists science in her service, the refractory hand of labor will always be taught docility.[2]

From the beginning of mechanization, with the invention of the Jacquard automatic loom, the self-acting mule, and other semi-automatic equipment, this management theme has echoed in the minds of inventors, including the earliest pioneers of computer automation. Thus Charles Babbage, the father of the modern computer, emphasized in his book on the economy of machinery and manufacturing (1832) that a "great advantage which we may derive from machinery is from the check which it affords against the inattention, the idleness, or the dishonesty of human agents."[3]

In our day too this obsession with management control continues to inform the dreams and designs of automation system developers. The earliest computer-controlled systems, created for use in the continuous process and mass-production industries, reflected this orientation, and this experience produced the technical experts who then went on to automate the critical metal-working industry. Metal-working, the guts of any machine-based and metal-based industrial economy, presented a challenge quite different from the others because it involved the small-batch manufacture of a wide variety of products using general-purpose machine tools. It lacked the high volume of products that would offset the great cost of specialized equipment, and, moreover, the wide variety of product alone defied the use of such special-purpose equipment. In addition, because of this requisite versatility, metal-working had always been very labour intensive and, as such, a bastion of worker militancy and a major obstacle to the mythical automatic factory. The task of the engineers, therefore, was both technical and social: to create a technology that would render universal machinery automatic without sacrificing its versatility (so-called "flexible" or programmable automation) and to create a technology that would enable management to discipline, deskill, and even circumvent and displace the machinist, thereby to gain complete control over production.

In the late 1940s control engineers at MIT (who had just completed a rolling mill control system designed to enable Bethlehem Steel management to eliminate "pacing" by workers) turned their "fertile genius" to the metal-working industry. The ultimate result of their efforts, "numerical control" (NC), reflected management's twofold objective and set the pattern for all subsequent development of what are now known as computer-aided manufacturing systems. As the very name suggests, control was and remains its essence, not just management control of machines but, through them, of machinists as well.

"With numerical control, there was a shift of control to management. The control over the machine was placed in the hands of management."

"I remember the fears that haunted industrial management in the 1950's. There was the fear of losing management control over a corporate operation that was becoming ever more complex and unmanageable. Numerical control is restoring control of shop operations to management."

"Numerical control is not a strictly metalworking technique; it is a philosophy of control."

"Numerical control has been defined in many ways. But perhaps the most significant definition is that [it] is a means for bringing decision-making in many manufacturing operations closer to management. Since decision-making at the machine tool has been removed from the operator and is now in the form of pulses on the control media, [NC] gives maximum control of the machine to management."

"There was little doubt in all cases that management fully intended to transfer as much planning and control from the shop-floor to the staff office as possible."

"The fundamental advantage of numerical control has been spelled out: it brings production control to the Engineering Department."

"In recent years, manufacturing industries in the U.S. have accelerated the move toward automating their operations. Factors that have motivated this move include the need to increase productivity, the high cost of labor, competition from abroad, and the desire for closer management control over production operations."[4]

The pattern is clear enough, the management motivation obvious from the time of the earliest factories to the present day. At the dawn of the industrial revolution, Andrew Ure boasted of how "in the resources of science, capitalists sought deliverance from [their] intolerable bondage" to the wit and will of the workforce.[5] The same is true at the dawn of this second, computer-based industrial revolution. Making explicit the management dream of progress without people, an engineer from the Arthur D. Little consulting company wrote excitedly to MIT after viewing an early demonstration of numerical control that the new technology signalled at long last "our emancipation from human workers."[6]

This enthusiasm on the part of engineers does not necessarily mean that they themselves have desired to eliminate or in any way injure workers. Such concepts are not part of the engineering vocabulary. But the engineers have objectives of their own, which neatly complement and innocently approximate those of management: they want to create an error-free system that will operate with a high degree of certainty in a manner perfectly faithful to the intentions of the designer. With this end in view, a "closed system," engineers need not deliberately seek, nor even be aware of, the management goals they so consistently serve.

Above all, engineers want to eliminate not particular human beings but the more abstract possibility of "human error." So they design systems that preclude as much as possible any human intervention. This is called "idiot-proofing." (In actual practice, it means designs that restrict intervention by all those except the engineers themselves and their managerial colleagues who, by their own estimate, could certainly never be considered idiots. Idiot-proofing, therefore, is the engineering equivalent to management's worker-proofing.) This engineering mentality betrays a rather cynical view of human beings (not to mention an elitist and derisive view of subordinates) in which any chance for human intervention (by workers) is negatively assumed to be a chance for error rather than, more positively, a chance for creativity, judgement, or enhancement. Like other engineering habits, it reflects engineers' privileged position in the industrial power structure. It is their relative power rather than their scientific training that enables and encourages them to design systems to be operated by "idiots." This is easily illustrated simply by situating engineers in a different, but equally familiar, context.

Suppose that an engineer (we'll say it is a man because the vast majority of engineers are men), decides to make a birthday present for his girlfriend. Since he is a mechanical engineer, he sets about to do what he does best, to build a machine. As her birthday approaches, he works day and night designing and perfecting this machine, which he hopes will embody the highest achievements of his art. Finally the birthday arrives and he presents his gift as a sign of his devotion, with a red ribbon tied around it. "Happy birthday darling," he says proudly, pointing to his elegant creation. "This is the most perfect machine I have ever designed. It's so perfect, in fact, that it can be operated by an idiot." His girlfriend is taken aback and their friendship is suddenly in doubt. Looking at him askance, she forcefully reminds him that she is not an idiot and demands that he return to his laboratory to redesign his machine for someone like her who is not an idiot. So the hapless engineer goes back to his labours and in desperation searches through the engineering textbooks and literature for clues as to how to proceed. Unfortunately, however, the assumption behind all of this learned material and all of his training and experience as well is that the best machines are those that are idiot-proof—that is, designed for managers wary of any worker involvement, who can force their subordinates to work like idiots. The love-struck engineer, therefore, does not even know where to begin to design a machine that would allow someone to intervene creatively as an equal.

The solution to his dilemma, of course, is to pretend that he is designing the machine for himself, since he surely doesn't consider himself an idiot. If he designed a machine that he himself would be operating, he would certainly leave ample room for his own later involvement in the process. Which is why, if there were a law requiring all machine designers to operate their own equipment for five or ten years after it was installed on the factory floor, there would be a revolution in engineering design.

Now, as it turns out, in the historical evolution of automated machine tools there have actually been attempts to do what the engineer's girlfriend wanted, but, despite their technical and economic promise, they all ultimately fell victim to the managerial obsession with control and its engineering counterpart, the quest for an error-free design.[7] The creators of all of these alternative designs shared a more respectful appreciation of the talents, knowledge, and resourcefulness of shop-floor workers and an understanding of their vital role in efficient, quality production. Thus men such as Eric Leaver, one of several inventors of record-playback control, F.P. Carruthers, designer of the Specialmatic control, and David Gossard, creator of the Analog Part Programming system—to name just a few—endeavoured to build machines for machinists rather than for idiots. The aim was to take advantage of the existing expertise not to reduce it through deskilling; to increase the reach and range of machinists, not to discipline them by transferring all decisions to management; to enlarge jobs, not to eliminate them in pursuit of the automatic factory.

Predictably, perhaps, but at any rate consistently, such alternative approaches remained stillborn. Their developers were perpetually plagued by a lack of funds and other forms of support, while promoters of the numerical-control approach enjoyed the sustained largesse of military and, later, industrial backers. Equipment purchasers, moreover, representing management, tended to reject the alternative designs at first sight since they were not compatible with management's overriding objective of control. Carruthers tried unsuccessfully in 1960 to win the support of the UAW; despite the validity of his claims that his system would save workers, jobs, and shop-floor prerogatives while NC would destroy both, the union failed to respond, no doubt reluctant to challenge such strictly management decisions. Finally, design engineers in general resisted this alternative approach as a matter of course, since it seemed to them messy, less predictable, and more open to human error.

Thus, NC became the dominant and, ultimately, the only technology for automating metal-working—the sole survivor and hence seemingly the best history had to offer, in the Darwinian view of technological progress. But, as we have now seen, this result was not the outcome of some natural selection by technical reason. Rather, it was the product of political selection by those powerful few seeking to retain and enlarge their social control, in league with those technicians who seek perfection in a world of idiots. In a machine shop in Lincoln, Nebraska, the ultimate fulfilment of these interwoven impulses, short of the totally automated workerless factory, was finally achieved. There, the NC equipment is run by a mentally handicapped operator with a maximum intelligence of a twelve-year-old. According to the *American Machinist*, this man was selected for the job "because his limitations afford him the level of patience and persistence to carefully watch his machine and the work that it produces." "His big plus," the shop's manager enthusiastically explained, "is that he will watch the machine go through each operation step by step. . . . He unloads every table exactly the way he has been taught, watches the [NC machine] operate, and then unloads. It's the kind of tedious work that some non-handicapped individuals might have difficulty coping with."[8]

The Military Mentality

A British machinist quipped recently that management is just a bad habit inherited from the military and the church, and there is much truth in the observation.[9] Whatever the church's role has been, the military mentality has certainly fuelled managerial obsessions from the start. And once having given rise to industrial management in the first place, the military has aided and abetted it ever since. The military's decisive impact on automation madness is a telling example.

The military has always played a central role in the technological development of U.S. industry, from mining and metallurgy to shipping and navigation, from interchangeable parts manufacture to scientific management.[10] As the army and navy have been the major movers in the past, the air force has led the way in our time (the Marines, apparently, have been otherwise occupied). If we just consider today's so-called high technology—electronics, computers, aerospace, cybernetics (automatic control), lasers—all are essentially military creations. When some of these war-generated technologies were brought together to automate the metal-working industry, the military was once again the driving force.

From the start in the late 1940s down to the present day, the air force has been and remains the major sponsor of industrial automation. With regard to numerical control, the air force underwrote the first several decades of research and development of both hardware and software, determined what the technology would ultimately look like by setting design specifications and criteria to meet military objectives, created an artificial market for the automated equipment by making itself the main customer and thereby generating demand, subsidized both machine-tool builders and industrial (primarily aerospace) users in the construction, purchase, and installation of the new equipment, and even paid them to learn how to run it.

Numerical control was just the beginning of air force involvement in the automation drive. The air force numerical-control project had global significance; on a recent visit to a locomotive factory in Prague, I was surprised to find the air force NC programming system in use even there. And before long, this single project had evolved into the more expansive Integrated Computer Aided Manufacturing Program. More recently, ICAM became the still more ambitious and diversified MANTECH (manufacturing technologies) programs, designed to promote the computer automated approach to manufacturing not only in industry but also in universities. "The Air-Force automation programs were established to force development of the technology," an ICAM program director explained several years ago. "Factories of the Future [the air force's latest boondoggle] are being designed to serve as models for U.S. industry in which computers and machines can be made to work together with little human intervention.[11]

The effects of this military involvement reflect the peculiar characteristics of the military world. First and most obvious is the military emphasis upon command, the quintessence of the authoritarian approach to organization. This means, essentially, that subordinates must do as they are told, with no ifs, ands, or buts; the intent is to eliminate wherever possible any human intervention between the command (by the superior) and the execution (by the subordinate). It is easy to understand the military emphasis upon automation, given its potential for eliminating such intermediate steps, as the ICAM director suggests. In the military outlook, an army of men behaving like machines is readily replaced by an army of machines. This command orientation neatly complements and powerfully reinforces the managerial obsession with control. If the business suit and the uniform are interchangeable in our day, so too are the minds that go with them.

The second characteristic of the military mentality is the focus upon performance above all else, reflecting the mission-oriented priorities of "combat readiness" or "national security." This fixation on performance renders all else secondary and fosters an industrial outlook that is more or less cost-indifferent. This explains the tendency toward waste, extravagance, and excess that marks so much military-sponsored effort, and it also explains why at so many U.S. factories today, while the American flag still flies overhead, Japanese machines are in use within. Preoccupied with meeting the exaggerated performance specifications of their No.1 customer, the military, U.S. equipment manufacturers have essentially priced themselves and designed themselves out of the domestic commercial market. (In 1978, the United States became a net importer of machine tools for the first time since the nineteenth century).[12]

Finally, the military's extravagance, however damaging to competitive industry, has proved extremely attractive to technical people, who are drawn to this "anything goes" atmosphere where they can try out their latest dreams. Fully a third of the nation's technical workforce go to work, directly or indirectly, for the military because it offers them the biggest technological playground. The money is the major incentive, of course, but not the only one.

Equally important are the technical enticements. With its nearly unlimited resources, the military offers technical people the often unique opportunity to work with state-of-the-art technologies at the cutting edge of development and the chance to dream expansively in pursuit of elegance and sophistication without regard to cost and other mundane practicalities. Moreover, with its rigidly defined chain of command and closely regulated environment, which guarantee a high degree of certainty and predictability, the military offers technical people almost laboratory conditions in which to try out their authoritarian designs. The military, in short, is able to indulge the collective enthusiasms and compulsions of technical people, which can be at once exhilarating and dangerous.

Enthusiasms and Compulsions

As everyone has experienced at one time or another, technical challenges can be highly seductive. It's not unusual to get emotionally involved when trying to make something work, whether the challenge is manual or intellectual. You skip dinner, ignore the calls of nature and other people, push on into the wee hours, driven, possessed, determined. There is a delight in it, a passion—and a blindness. You can hardly tolerate interruption or delay, much less interference, and you get so you would almost kill to get the damn thing to work (which is what so many technical people eventually do; they put their talents at the disposal of the military—which, after all, is in the killing business—in order to be able to follow through on their compulsions). Of course, such emotional enthusiasm is the wellspring of creativity and can often be inspiring and enriching. But when it is indulged beyond reason, in defiance not only of personal health but also of the larger social welfare, it becomes madness. Let us take a closer look at some of these enthusiasms and compulsions, the ones that underlie automation madness.

During nearly a decade of teaching at MIT, a high-powered engineering school, and especially during the course of my study of industrial automation, I became increasingly convinced that there were collective psychological forces at work that defied conventional political and economic analysis. It seemed to me that the people who were caught up in the emotional excitement of automation shared not only a set of technical interests and fantasies but also a common underlying compulsion. They were the drivers of the automation advance, yes, but there was also something driving them, something shaping their perceptions of reality and their image of an ideal world. For several years now I have been trying to understand what this is about, and here I can only summarize my speculations.

To begin with, there is the shared ideal of a world without people, an image that affects the way these people view the activities around them; that is, their imagination distorts their perception. Consider, for example, the perceptions of Andrew Ure, the nineteenth-century authority on manufacturing. When Ure examined early textile factories, this is what he saw: "I conceive that this title — Factory — in its strictest sense, involves the idea of a vast automaton composed of various mechanical and intellectual organs, acting in uninterrupted concert for the production of a common object, all of them being subordinated to a self-regulating moving force."[13]

Given this passage from Ure's *Philosophy of Manufactures*, it is easy to forget that these early factories were teeming with people, people who magically disappear in Ure's description. This is a common (mis)perception of technical enthusiasts. Nearly a century and a half later, to take a more contemporary example, Dr. C.C. Hurd, Director of Applied Science at IBM, offered a strikingly similar observation: "It seems to me that the most useful analogy which I can see for the assembly line is that the assembly line — or, more generally, a complete production line — is like a computing machine."[14]

Whereas Ure entered a factory full of workers and saw only a vast automaton, a self-acting machine, so Hurd (in the manner of his likeminded ancestor Charles Babbage) saw only a computer. The power of abstraction of such men blinded them to the actual human realities of production. And when these realities rudely but invariably interfere in their reveries, they are viewed with contempt and arrogant impatience. For technical people absorbed by such imaginings, the ideal often becomes more real than reality itself: a fantasy of a perfectly ordered universe to which the world of people must be forever adjusted. The attractions of this idealized world of machines and computers are clear enough; this is above all a clean world, controlled, predictable. But there is more to it: if such a vision fulfils a deepseated desire for order, it also satisfies an enchantment with things that are at once animated and artificial, almost life-like in their autonomy, on the one hand, and yet under nearly complete (albeit remote) control, on the other. For these dreamers, there is sheer delight in such a spectacle. Where does this enthusiasm, this intrinsic fascination with automation come from?

One likely clue is the fact that this fascination with automation, an ancient obsession, has always been a peculiarly male preoccupation and remains so today. There is no evidence suggesting that women have shared this keen interest in what might be termed the artificial creation of life. Since ancient times, men alone have sought to play God by constructing mechanical devices that appear to imitate Nature. This artificial act of creation has been matched by equally common artificial acts of procreation, of men playing not God but woman—presumably to compensate for the male inability to give birth to children (womb envy).[15] In certain societies, for example, it is common for men to engage in what are known as couvade rituals. While the women go off into the woods to have their children, the men gather in the village centre to perform their often elaborate rituals in which they act out the birth process. In this way, they symbolically appropriate this life-giving act for themselves and place themselves at the centre of reproduction rather than at the periphery, where they actually are. This charade meanwhile reduces the role of the women, symbolically rendering them irrelevant and invisible.

Similarly, throughout the Middle Ages and later, men of learning devoted themselves to discovering the secret of life and to devising methods for fathering offspring without the aid of women. (Such mythical children were known as "humunculi.") For example, Paracelsus, who was probably the leading medical figure of the sixteenth century, offered a recipe for growing a humunculus in warm horse manure by mixing human semen with the blood of animals and waiting for forty days. The resulting child, Paracelsus promised, would be smaller than the average human but, being the child of the father alone, would be of superior intelligence.[16] With modern genetic engineering, this male drive to create life without women continues in our own day.

> None but those who have experienced them can conceive of the enticements of science. No one can conceive the variety of feelings which bore me onwards, like a hurricane, in the first enthusiasm of success. Life and death appeared to me ideal bounds, which I should just break through, and pour a torrent of light into our dark world. A new species would bless me as its creator and source. Many happy and excellent natures would owe their being to me. No father could claim the gratitude of his child so completely as I should deserve theirs.[17]

This familiar passage is from Mary Shelley's classic tale *Frankenstein*. Shelley knew very well that she was describing a male obsession—it is the theme of the novel. Imagine for a moment what would happen if Dr. Frankenstein, the possessed creator of the monster, were a woman. The story would no longer make sense; the demonic frenzy, the insane drive, would disappear to be replaced by the natural process of childbirth. Women can create life without resorting to such scientific contrivance. It is only men who have this strange compulsion, to create life in artificial ways without women.

Could it be, then, that automation madness, the drive to artificially create life-like machines, derives from this same source? Could it be that automata are a means by which men try to compensate for their biological incapacity to give birth, a means at the same time for artificially achieving some continuity and immortality? Consider the words of one automation engineer: "The automatic factory will not only be larger and higher, but it will penetrate the ground much deeper. Furthermore I believe it win be nuclear hardened to survive a nuclear attack."[18]

If there is any truth to this speculation — and the vast majority of automation engineers are men, after all — then perhaps the guiding ideal in all this effort is not only a world without people but also, more particularly, a *world without women*. That is, perhaps the automation enterprise is the couvade ritual for modern industrial society. Think about this the next time you watch engineers with their machines. I recently visited the National Bureau of Standards, which has under way an extensive robot manufacturing development program. I observed that all the engineers and technicians at work in the model automatic factory were men, and their relationship with their custom-made creations struck me as peculiarly intimate.

A decade ago the British government commissioned a study of artificial intelligence (a technical field central to automata design). The final report of this study, authored by the eminent Cambridge University mathematician Sir James Lighthill, contains this provocative passage.

> It has sometimes been argued that part of the stimulus to laborious male activity in creative fields of work, including pure science, is the urge to compensate for lack of the female capability of giving birth to children. If this were true, then building robots might indeed be seen as the ideal compensation. The view to which this author has tentatively... come is that a relationship which may be called pseudomaternal comes into play between a robot and its builder.[19]

Of course, again, all of this is just speculation. But it seems that there is something going on here that warrants serious attention. For whatever it is that is driving men to automate, it is also driving us all in the same direction, a direction in which we might not really want to be heading, or at least not at the current velocity. What is the goal? What is the hurry? There is something out of control here, something almost transcendent (not to mention socially irresponsible). "The automatic factory is like the Holy Grail—something you approach but never reach," the president of Automatix, Inc. solemnly observed in 1983.[20] *Fortune* magazine put it even better thirty years earlier, just as the automation advance began to accelerate: "In the nature of things, man will create an automatic factory as he climbs Mt. Everest and aspires to reach the moon, for reasons no one has ever clearly expressed. Except that he is a man."[21]

BEHIND THE ECONOMIC SCREEN

By now I am sure the reader is thinking, okay, so these technical enthusiasts are a little nuts, but so what? Fortunately, our ideology tells us, their excesses are corrected by the economic screen of automatic technological progress—by the no-nonsense economic rationality of profit-seeking businessmen and by the ever-dependable self-regulating mechanism of the market. Would that it were true. For here too reality belies the Darwinian assumptions of our mythology.

THE ALL-TOO-HUMAN BUSINESSMAN

Our ideological image of the businessman is a caricature of a hardworking, practical-minded "economic man," guided by sober cost-accounting and the pressures of competition and supply and demand, and intent above all upon making a buck. This stark abstract image, too inhuman to be real, is portrayed by the press, by public relations agents, and by businessmen themselves (not to mention by economists of all political persuasions) because it conveys the impression that business itself is abstract and therefore beyond our control, an objective, inevitable force in our lives rather than merely a mad scramble of greedy and familiar people. And this objective appearance only reinforces our idea of that automatic destiny we call progress.

Businessmen, we are reminded almost daily, have a job to do; they have no time for dreamers. Businessmen are straightforward, down-to-earth, simple to understand. They are predictable. They are ruthless in their blind pursuit of profit, so we can count on them. This is why we have such confidence in the economic screen—this sublime self-interest takes on a life of its own. Just as we have believed that technical people are merely agents of an abstract autonomous technical rationality, so here we assume that businessmen are simply agents of an abstract and autonomous economic rationality. The engineer evaluates the machine by asking, will it make widgets? The businessman looks again and asks, will it make money? Given their dedication, we can't lose; we can't fail to get a technology that is not only technically but also economically viable.

But, alas, as we have now seen in the case of the technical people, appearances are deceiving. For the businessman is more human than our caricature allows, more like the real technician — and the rest of us. He too has dreams and delusions, enchantments and enthusiasms, flights of fantasy: "U.S. companies are on the verge of achieving a dream . . . [of] manufacturing enterprises where push button factories and executive suites, no matter how physically remote become part of the same computerized factory."[22] Of course, businessmen want to make a profit. They believe that their actions will have that result, and they justify those actions always in economic terms. But this is far from the whole story. Let us take a closer look at what moves the businessman and, through him, the advance of automation technology.

For one thing, justifications are not the same as motivations. In reality, the former tend merely to obscure the latter and serve instead as rationalizations for actions taken for unstated reasons. Despite the authoritative appearance of such economic calculation — with every cost estimated down to the last decimal point — there is typically less there than meets the eye. I well remember my own initiation into this murky world of industrial economics.

While teaching at MIT, I was invited to take part in a study of automation at General Motors (which GM subsequently cancelled to prevent site visits or union involvement). At the first meeting on the project, I found myself in a room full of seasoned industrial experts — engineers, economists, labour relations people. The discussion began with the obvious questions. Why was GM automating? Why does anyone automate? And the response to the questions was immediate and nearly unanimous: people automate to make money, to increase profits. As everyone else moved on to the next issue I sat there in sceptical silence. It seemed too simple, somehow. Finally I got up the temerity to speak to these experts, and I asked if any of them had evidence that automation was profitable or even cost-effective. Everyone looked at everyone else. They had none. So, I asked, aren't we just making an assumption here? We believe businesses automate to make money, yet we don't know if they actually do. (I later found some evidence of this but also evidence of the opposite, that they lose money. The results to date are ambiguous, as we will see below).

I pointed out that people continue to automate, but it is only conjecture that they are actually making money or even that they are doing it just to make money. It's true they might know something the assembled experts did not know, but they might also merely be acting on faith too, on the belief that automation is profitable — just like the experts. On the other hand, they might be automating for altogether different reasons. Economists argue that businessmen would not automate if it wasn't profitable; they automate, so it must be profitable. But this is just logic. Where is the evidence? What is really going on in the minds of businessmen? I thought at the time that my questioning would give the experts pause and stimulate some reflection and investigation. I failed then to understand the convenience of an ideology that gives the answers without such effort. Undaunted by the challenge, my colleagues simply went on without me.

A year or so later I ran across an interesting series of articles by a young Harvard economist, and my scepticism was fuelled.[23] The subject was what is known as relative factor analysis. According to neoclassical economics, businessmen decide whether or not to invest in machinery by comparing its cost with the costs of labour. According to this theory, if the cost of machinery is less than the cost of labour, they will invest in machinery, and if the cost of machinery is more than the cost of labour, they will stick with labour. This intrepid young economist undertook to test this theory in the field by conducting a survey of some sixty factories in the New England area. In each case he identified and talked with the people who actually made such purchasing decisions and tried to find out how they did it.

He discovered, first of all, that the majority of these people had technical backgrounds; they were engineers. He also found out that their actual purchasing behaviour differed from what the theory suggested. When the cost of machinery was lower than that of labour, they bought machines, but when the cost of machines was higher than that of labour, they still bought machines (and sometimes fudged the justification accordingly). The economist concluded that there was a bias in favour of machinery (or against labour) on the part of these technically trained functionaries. Their enthusiasm for machinery was the major determining factor, not careful relative factor analysis.

As my own study on automation got under way, I started visiting factories myself. Before long I had got used to the unexpected. I remember talking with a man who installed a particular type of numerical-control equipment. He told me a strange story of how in one shop he noticed a row of assorted castings lined up against the wall next to where he was installing the new machine. After asking around, he learned that these machining jobs had been used to justify the purchase of his machine. He was astonished, because his machine could not machine castings; it was a turret punch press that worked only on sheet metal.

In other shops, where the jobs at least matched the capability of the equipment, I learned that, while managers consistently boasted about how the new equipment increased productivity, they never seemed to have any hard evidence. I found out also that productivity is itself a slippery concept, hard to define much less measure, and also that so-called "creative accounting" among divisions of a company spreads the data around in such a way as to make it almost impossible even to make an educated guess. More amazing still, I discovered that there were almost never any post-audits done on equipment, to assess after the fact whether or not expected benefits (used to justify purchase) had actually been realized; apparently few people want to learn from or even to document their errors, so, once a machine is put in concrete, it usually stays there — and so does the next one. And one more thing: I learned from my colleagues that productivity was not necessarily the critical fact in assessing machines anyway. By means of creative accounting and sophisticated use of the tax laws, machines can mysteriously make money for their owners even if they don't work or are never used. Little did I ever suspect that machinery could be profitable as furniture.

Similarly, and finally, I visited the National Bureau of Standards to attend a demonstration of a newly developed computer-aided manufacturing system. Along with about fifty or so businessmen from firms around the country, I was treated to a show-and-tell and then ushered into an auditorium for a film and discussion. After about thirty minutes, I was the only person there who asked anything like a "bottom line" question. What data did they have for us, I asked, pertaining to the expected cost effectiveness of this fancy equipment? The NBS official politely dismissed the question as not being a matter within his purview. The guy sitting next to me — I believe he was from General Electric — leaned over and whispered, "We don't do that any more. We just have to get into this with both feet no matter what, we don't have a choice." At that point, I started looking for the free lunch.

After all of this experience, I began to believe that, rhetoric and theory and ideology aside, careful economic consideration of technological development was no longer the order of the day — if it ever was. So I started to look elsewhere for clues about what was driving automation, what was really motivating all of these supposedly cautious, calculating businessmen. It seemed to me by now that there was a myriad of motivations — political, cultural, psychological. Some of these were couched in economic jargon, and all were routinely justified in economic terms, but they actually had little to do with economics. Instead, they resembled those more human and familiar obsessions, enthusiasms, and compulsions already described above. Although the vaunted economic rationality of businessmen sometimes comes into play, more often than we suspect these other impulses underlie the decisions on technology. And only after the choice has been made, only after the equipment has been installed, is there any serious effort to render it economically viable — with mixed results, as we will see, and usually at public expense.

As the Harvard economist's study indicated, the promoters of new technology within companies are typically people with technical backgrounds and the enthusiasm for technology that goes with it. In 1981 Donald Garwin of Arthur D. Little, Inc. conducted a study of automated flexible manufacturing systems (FMS) and found: "Management is usually sold on the idea to use such systems by an engineer in the company who is enthusiastic about the technology. Cost justifications play a secondary role. The more sophisticated and fascinating a machine is, the less management is likely to quarrel over dollars."[24]

Top management, moreover, have enthusiasms of their own. Each year, for example, there are major machine-tool shows around the country, where manufacturers display and demonstrate their latest wares for prospective buyers. These shows are extravaganzas, like boat shows or automobile shows where people go to view the latest, fanciest yachts or sports cars. The displays are made as attractive as possible.

Typically, as at boat shows or auto shows, female models are posed provocatively beside the equipment, which is usually painted in primary colours to catch the eye — bright red or yellow or blue. (Meanwhile, behind the scenes, harried technicians struggle desperately to get the unreliable equipment to operate flawlessly for the periodic demonstrations.) Managers come to these shows on company time, to relax, socialize, and catch up on progress. So on one such occasion, a manager is having a good time joking with the affable hostesses at an exhibit. He's in a receptive mood when he starts walking through the lobby, and suddenly he comes upon a really sexy automatic robot machining centre, a red one, and he says to himself, God, I'd love to have one of these in my shop.

So he races back to his office and instructs his staff to get one of those machining centres — a red one. Some subordinate engineer is assigned the task of justifying the purchase. Being new to the game, this subordinate does a careful and thorough job and concludes that the company cannot afford to make the purchase. Even though he himself is enthusiastic about it, his estimate shows that it would be too costly and unreliable. He sends his report upstairs, and only minutes later, before he knows what happened, his superior rushes in to chew him out. "What's going on here, what are you, a saboteur? The boss wants one of these machines!" So the young engineer returns to his desk and reworks the justification so that it comes out the way the boss wants it. (As far as I can tell, management fudges on justifications even more than students cheat on exams.) Before too long, the shiny red machining centre appears on the shop floor — symbol of the boss's progressive outlook — and if the company is large enough (or subsidized enough) to absorb the initial cost and subsequent downtime, it stays in business.

This kind of thing happens frequently, but economists rarely talk about it, because they are only concerned with so-called objective factors, not the real human ones. Joseph Engelberger, founding president of Unimation, Inc., the first major U.S. robot manufacturer, knows better. As a super salesman of industrial equipment he well understands what sells machines, and he is candid about the enthusiasms of his customers and their preoccupation with such intangibles as social status. He remarked to the press, "I don't think a guy will be able to go into his country club if he doesn't have a CAD/CAM [computer-aided manufacturing and design system] system in his factory. He's got to be able to talk about his CAD/CAM system as he tees off on the third tee — or he will be embarrassed."[25]

In addition to the concerns of individual managers there is a collective phenomenon at work as well, a sort of herd instinct, in which they all get caught up together. It's something like a run on a bank or the stock market. Someone starts it and before long everyone is doing it out of desperation. Managers feel they must automate because "everyone's doing it," out of fear that they will be undone by more up-to-date competitors (a paranoia encouraged by equipment vendors). There is this vague belief that the drive to automate is inevitable, unavoidable, and this belief becomes a self-fulfilling prophesy. In the stampede, meanwhile, there is very little sober analysis of costs and benefits. "Everybody and his brother believes that FMS [flexible manufacturing systems] is the only way to fly," the trade magazine *Iron Age* reported. "Yet, there isn't a single FMS in the U.S. that operates the way it was intended to." One Boston University business school professor told the *Wall Street Journal*, "Companies are buying equipment helter skelter without thinking about how they want to use it."[26] In short, this trance that we are all in—the feeling that there is some inevitable force called technological progress and you have to hop on or get run over—they're in it too.

This ideology, as we now see, begs all of the questions and avoids all of the answers about the reality of technological progress. Leaving aside the human and social dimension, the actual political, cultural, and psychological factors at work, the ideology allows for no serious scrutiny of actual motivations and real returns. Insofar as it does this, it serves a purpose—to camouflage, obscure, dignify, and ratify the actions of those in power.

In 1983 I experienced this at first hand. I was invited to testify before Congress on the plight of the U.S. machine-tool industry and industrial policy in general (see chapter 6) I was put on a panel with industrial executives from companies that manufacture and use machine tools. They had come to Congress to ask for protection against foreign imports; maintaining that they themselves were doing a good, honest job, they decried the unfair practices of their foreign competitors. When it was my turn to speak I presented a short history of the U.S. machine-tool industry, describing the human and social ways in which it had undermined its own position: the role of the military, the enthusiasms of technical experts, the managerial obsession with control, and the various excesses and foibles. I pointed out to the Congressional Committee that these businessmen were not quite the economic rationalists they pretended to be; rather, I argued, they were people just like the rest of us, with strengths and weaknesses too.

Before I was halfway through my prepared statement, the executives on the panel became visibly upset. One of them abruptly demanded the opportunity to rebut my position point for point in writing (this was granted, but the rebuttal never materialized). Later a staff person who had helped set up the hearings told me that one of the executives confided to her that he would have liked to punch me out right then and there.

What is the reason for such hostility? The answer is simple. If you talk about these executives in the way you talk about anyone else, you blow their self-righteous cover, undermine the disarming dignity of the objective economics in which they conveniently envelop themselves, and violate the academic and popular myth that they really know what they are doing. Leave the driving to us, they confidently shout from the cab of the locomotive of progress, as they head toward the next mountain turn. Is this really the safest way to travel?

THE MARKET MIRAGE

If you can't trust the technical people and you can't trust the businessman, whom or what can you trust to keep technological progress on course? Happily, there's still the market, that mysterious yet infallible mechanism that magically makes everything work out in the end. Just as it miraculously transforms the individual pursuit of self-interest into the larger social good, so it consistently corrects for the excesses and errors of individual businessmen by forcing them into bankruptcy and out of the picture. Only the sober, smart, and savvy survive and thus, finally, in this competitive court of last resort, our Darwinian assumptions of natural selection are upheld. Not quite.

The convenient fiction of the market was a nineteenth-century propaganda invention created by a upwardly mobile bourgeoisie to challenge the economic power of the state and thereby extend the range of their exploitation.[27] In reality, the "free" market has never truly existed, because businessmen have always used all the political power at their disposal to influence events in their own interests: they used the state to create the "free market" in the first place by doing away with regulations protecting workers and consumers; they enacted all sorts of protective devices for themselves, from state-chartered and subsidized corporations and tax incentives to military support of enterprise and, of course, tariffs.

The same is true today, when the role of government in the economy is greater than ever before. The supposedly self-regulating mechanism of the competitive market is easily overwhelmed by the power of the state as both underwriter of enterprise and largest customer. In the case of automation, as we have seen, the state, especially the military, has played a central role. Not only has it subsidized extravagant developments that the market could not or refused to bear, but it also absorbed excessive costs and thereby kept afloat those competitors who would otherwise have sunk. As one air force official candidly observed:

> We have contractors with divisions set up just to get Air Force projects. We're keeping them alive. People are automating for automation's sake in several cases. There is no good reason, there is no good justification — and in fact it may be detrimental. We work with companies whose job it is to implement these advanced technologies, and if they can get a project from the Air Force, regardless of its real payback, they keep in business.[28]

It is thus no accident, for example, that the U.S. machine-tool builders trade association moved its headquarters from the midwest centre of the industry to Washington, D.C., home of its major customer, the Department of Defense. Nor is it an accident that the defence-related industries are the ones with the most automation. These industries, moreover, are expanding along with the military automation programs, as more and more businesses rush to this state-supported sanctuary to escape the unpredictable vicissitudes of the market. At the same time, the military automation programs are today being matched by those of civilian agencies such as the Department of Commerce, the National Science Foundation, and others. All have now become the publicly funded pushers of automation madness, charting a course and promoting a pace that no self-adjusting market, had it existed, would ever have tolerated.

Where the state fails to provide safety from competitors, monopoly succeeds. The economic power of gigantic multinational corporations, some of which exceed the scale of governments, allows managers to carry costs, and conceal costs, that would cripple other firms. Their sheer economic (and thus political) muscle enables them to corner markets, intimidate or "acquire" competitors, and thereby distort beyond measure the real costs of doing business. And the relationship between corporate profit and economic production is becoming more incidental every day. The corporate automation drive is just one case in point. Not surprisingly, the giant firms are the leaders in this drive, and it is difficult if not impossible to evaluate their returns. General Electric is a prime example (it is also a major, and heavily subsidized, defence contractor, like many giant multinational manufacturing companies).

G.E. decided several years ago to become the "world supermarket" for automation equipment, the largest supplier of such industrial machinery. With this strategy in place, G.E. accelerated the introduction of its automated equipment within its own factories. At each location (Louisville, Erie, Schenectady, Lynn) and in each product division (appliances, locomotives, turbines, aircraft engines) the company insisted that it had to automate to stay competitive, despite the loss of jobs. But how much of this effort is really a marketing strategy to sell its equipment to other companies? By making some of its own plants showcases of automation (and absorbing the costs elsewhere in the corporation) G.E. kills two birds with one stone. The company intimidates the unions into concessions and acquiescence to job loss, while at the same time it holds up these shiny robotized plants as examples of the factory of the future in order to sell more equipment. The company's powerful position in all of these markets, its ability to shift costs internally, and its ample state support all guarantee its continued survival and prosperity—despite the half-truths about competition presented to the unions at contract time, and whatever the actual costs and benefits of automation.

Thus, the market panacea turns out to be just one more mirage that evaporates upon closer inspection. No automatic guarantor of economically sound technological progress, it is instead yet another ideological camouflage for political power. Perhaps it is time now to leave Darwinism to biology, where it belongs, and to start looking at this important matter of technological progress more critically, because it has serious consequences for us all. Having overcome the first half of the mythology of automatic progress by examining more closely the human and social drives behind technological development itself, let us turn now to the other half, to examine more closely the consequences of these drives—where they have led and are still leading us. According to our inherited ideology of progress, automatic technological progress leads automatically to social progress. But if, as we have seen, there is really nothing automatic about technological progress itself, what can we expect in the way of social progress? Could this turn out to be a story without a happy ending?

NOTES

1 In the late 1960s the management at G.E.'s Lynn, Massachusetts, plant was having technical difficulties with newly installed numerical-control lathes and blamed it on worker sabotage. In reality the workers were trying to correct for problematic equipment by manual intervention, but were hampered by management's insistence that machines could run by themselves, and that the workers were only "button-pushers" or "monkeys." After considerable conflict, G.E. introduced a quality of worklife program (a prototype of those later introduced in the auto industry), which gave workers much more control over the machines and the production process and eliminated foremen. Before long, by all indicators, the program was succeeding: machine use, output, and product quality went up; scrap rate, machine downtime, worker absenteeism, and turnover went down; and conflict on the floor dropped off considerably. Yet, little more than a year into the program — following a union demand that it be extended throughout the shop and into other G.E. locations — top management abolished the program out of fear of losing control over the workforce. Clearly, the company was willing to sacrifice gains in technical and economic efficiency in order to regain and ensure management control. (See the final chapter, "Who's Running the Shop," in my *Forces of Production* (New York: Knopf 1984).
2 Andrew Ure, *Philosophy of Manufactures* (Burt Franklin, 1967), pp.336-68.
3 Charles Babbage, *On the Economy of Machinery and Manufactures* (New York, 1963), p.54.
4 Earl Troup, personal interview, 1977; Willard Rockwell, Chairman of the Board, North American Rockwell, address to Western Metal and Tool Exposition, March 11, 1968 (North American Rockwell Corp., 1968); *American Machinist*, October 25, 1954; Nils O. Olesten, "Stepping Stones to NC," *Automation*, June 1961; Earl Lundgren, "Effects of NC on Organizational Structure," *Automation*, January 1964; *Iron Age*, August 30, 1976; *Industrial Engineering*, April 1984, p.50.
5 Ure, *Philosophy of Manufactures*, p.369.
6 Alan A. Smith to J.O. McDonough, September 18, 1952, NC Project Files, MIT Archives.
7 A series of such efforts are described in great detail in the chapter "The Road Not Taken" of Noble, *Forces of Production*.
8 "Special Reasons to Hire the Handicapped," *American Machinist*, July 1979.
9 British machinist, quoted by Mike Cooley in talk at MIT, winter 1979.
10 For a sustained examination of the role of the military in technological development, see Merritt Roe Smith, *Military Enterprise and Technological Change* (Cambridge, Mass: MIT Press, 1985).
11 Dennis Winosky, quoted in Jerry Mayfield, "Factory of the Future Researched," *Aviation Week and Space Technology*, March 5, 1979, pp.35-37.
12 For a penetrating examination of the negative effect of the military influence on commercial enterprise, see Seymour Melman, *Profits Without Production* (New York: Alfred A. Knopf, 1983).
13 Ure, *Philosophy of Manufactures*, p.1.
14 C.C. Hurd, quoted in "The Automatic Factory," *Fortune*, October 1953.
15 For more on the long history of automata, see John Cohen, *Human Robots in Myth and Science* (New York: A.S. Barnes and Co., 1967).
16 Paracelsus, cited in Cohen, *Human Robots*, p.44.
17 Mary Shelley, *Frankenstein* (New York: Pyramid Books, 1957), pp.44-45.
18 Frank McCarty (Raytheon Corp. executive), quoted in Daniel B. Dallas, "The Advent of the Automatic Factory," *Manufacturing Engineering*, November 1980.

19 Sir James Lighthill, *Artificial Intelligence* (London: Science Research Council, April 1983).
20 The president of Automatix Inc. quoted in Dallas, "The Advent of the Automatic Factory."
21 "The Automatic Factory," *Fortune*, October 1953, p.195.
22 *Business Week*, December 13, 1982.
23 Michael Piore, "The Impact of Labor Market upon the Design and Selection of Production Techniques within the Manufacturing Plant," *Quarterly Journal of Economics*, vol. 82 (1968).
24 Garwin cited in Vera Ketelboeter, "Where Is Automation in Manufacturing Headed?" *Science for the People*, November/December 1981.
25 Engelberger, quoted in Gene Bylinsky, "A New Industrial Revolution Is on the Way," *Fortune*, October 5, 1981, p.114.
26 *Iron Age*, cited in Tom Schlesinger, *Our Own Worst Enemy* (New Market, Tenn.: Highlander Center, 1983); Stephen Rosenthal, quoted in "High Tech Track," *Wall Street Journal*, April 11, 1983.
27 For an enlightening history of the idea of the market, see Karl Polanyi, *The Great Transformation* (Boston: Beacon Press, 1944).
28 Gordon Meyer, quoted in Schlesinger, *Our Own Worst Enemy*.

CHAPTER 5
A SECOND LOOK AT SOCIAL PROGRESS

In 1952, at the dawn of the so-called age of automation, a Pratt and Whitney engineer waxed eloquent about its promise. A pioneer of automation himself, he used the central metaphor of cybernetics to express his quasi-religious faith in the automatic beneficence of technological progress: "I don't think we are consciously trying to ease the burden of our workers, nor consciously to improve the standard of living. These things take care of themselves. They have a feedback of their own that closes the loop automatically."[1]

This faith takes many forms in our culture and is rarely if ever articulated with any clarity, much less logical rigour. But in outline this ill-defined faith could be called the beneficent circle of prosperity. It goes like this—people with money are offered incentives (the chance to make more money) to urge them to invest in a new, improved plant and equipment (so-called innovation). This innovation automatically yields increased productivity and, hence, lower costs and prices, which results in greater competitiveness. Finally, this enhanced competitiveness necessarily brings about what Adam Smith, the great eighteenth-century philosopher of capitalism, called the "wealth of nations": economic growth, jobs, cheap and plentiful commodities, in short, prosperity.

Let us now take a closer look at this magic circle, to examine these assumptions on their own terms. For when we do, we will see that the causal chain is, in reality, ambiguous at each link, and the end result is not exactly what we blithely assume it must be.

What Goes Around Comes Around

To begin with, the assumption that rich people will invest in new means of production if given sufficient lucrative incentives presupposes that they would not do so voluntarily without such inducements. As such, it is itself a tacit recognition of the inadequacy of the market as a stimulus to development. Furthermore, it rests upon the prior assumption that people simply follow their pocketbooks, which, while largely true, is not, as we have seen, the whole truth. More to the point here, to the extent that people do strive primarily for the highest return on their investment, there is no guarantee that they will invest in new means of production if other, more profitable (even given incentives) routes are available.

It certainly appears to be the case in our own day, judging by patterns of investment, that however vital production remains to any economy, it has become relatively less attractive as a way to make money. In the past, production was the point of entry for ambitious capitalists closed to more established avenues of power and wealth. Today, this historical connection between capitalism and production appears to be fading as an increasing proportion of investment is diverted into non-productive areas of the economy: real estate, or financial speculation, for instance. Indeed, there seems to be a scramble among capitalists to get out of the messy, troublesome business of actually producing something for society whenever they have an opportunity, a trend toward disinvestment that is leaving a trail of debris — closed plants, idle workers, ghost towns — in its wake. In an age when oil companies invest in circuses, manufacturers invest in real estate, and the steel industry deliberately and openly abandons the production of steel, there is good reason to reconsider this first link between investment and innovation.[2]

The second causal link — between innovation and productivity — is more difficult to assess, but judging from the case of automation it too appears to be ambiguous at best, nothing solid enough to rely on. When investment does in fact generate innovation, does such innovation necessarily yield greater productivity? The assumption here is that the return of profits to the investor will be matched by more and cheaper goods for society. This assumption, of course, is the cornerstone of apologies for capitalism, its central tenet of legitimation. But today even the business press has begun to back away from this claim. After conducting a poll of industry executives on trends in automation, *Business Week* concluded, "There is a heavy backing for capital investment in a variety of labor-saving technologies that are designed to fatten profits without necessarily adding to productive output."[3]

But few are able to confront, much less draw the correct implications from, such ideologically disorienting disclaimers. Thus, efforts to document productivity gains continue to abound, despite the fact that the concept of productivity is hard to define and the reality equally difficult to measure. For one thing, there is very little hard information available.

Even my more optimistic colleagues at MIT concluded from their preliminary investigation of industrial automation: "After-the-fact analysis of the actual economic impact of a process of automation is rarely carried out; the result is a loss of data necessary to inform subsequent decisions about automation."[4] After abandoning my own attempts to get hard data I resorted to the less "scientific" approach of asking people in the factories I visited. This proved problematic, too. According to the managers, every new machine increased productivity, often dramatically. After a while the plausibility of this too consistent claim began to wear thin, especially given the fact that the shop-floor people who worked with the equipment invariably told a different tale, of downtime and disasters. The reality of productivity, however one defined it, remained ambiguous; in recent years, even the business press has begun to acknowledge this. "The results are mixed," the *Wall Street Journal* reported in 1980. "As technology soars, users struggle with the transition and unsuitable machines; computerized equipment often doesn't work the way it's supposed to, the new equipment is more fragile than the old-fashioned equipment . . . and problems with the software used to run the equipment are even more prevalent." As Thomas Gunn of Arthur D. Little put it, companies "get the biggest, fastest, sexiest robot, when the plain truth is that in most cases a very simple piece of equipment could do the job," and they "don't so much make mistakes as learn that it's going to take two or three times as much money and time as they thought to get the system working."[5]

Economists typically respond to such news of technical unreliability and economic uncertainty with calm confidence, arguing that industry is now just going through the learning stage and that, after a while, the expected productivity gains will be realized. But this wishful thinking—which disarms critics and forever defers judgement—is perhaps too sanguine and smug, as the experience of the banking industry, which has had more time to move along the "learning curve," now suggests. A study of the economics of innovation in this area reported by *Computing Canada* on May 16, 1985, found, "Senior executives of banks are generally disappointed in the return on their investments in technology." It stated: "The inability to use technology to achieve lasting competitive advantages, and the failure to achieve expected economic returns through reduced operating costs, were among reasons given by senior management of 200 major banks and financial institutions in 26 countries surveyed by the management consulting firm."[6]

At least some inside observers have begun to acknowledge a similar disparity between religion and reality in the industrial automation experience. Henry Miley, retired air force general and director of the American Defense Preparedness Association, expressed his concerns about the economic returns on military investment in automation in testimony before the House of Representatives in 1980.

> What concerns me is that when I get up and raise my Yankee voice and say, can I go out to some factory and put my hands on an item that is being produced more cheaply now than it was five years ago because of the [air force automation programs], I get kind of a confused answer. When I ask the bottom line question, is the [product] now cheaper than it was two years ago because [of the air force program], the answer was, well, no.[7]

Apparently, the ambiguity of results reflects the overriding performance imperatives of the military, not to mention the enthusiasms of technical people and the preoccupation of management with control and the fast hustle. But since these factors pervade those very industries that are undergoing automation, there is certainly reason to generalize from General Miley's assessment. Has anyone noticed any decline in prices lately?

Given these mixed results, one might expect any rational person to abandon the ready assumption that all innovation increases productivity and to seek more sober assessment. Yet any such effort to do a careful and critical assessment, to try to separate the wheat from the chaff, immediately invites the charge of "Luddism," enemy of progress. Ironically, in our day, the demand for greater productivity has become a revolutionary slogan rather than a paean to capitalism, because it exposes the soft, ambiguous underside of our seemingly authoritative economic justifications for domination.

For example, most analysts and industry accountants measure productivity as output per person-hour, that is, product per unit of labour time, where labour means hourly production (direct) labour and time is hours on the job. (The engineers I have talked to typically use a standard number to estimate the cost of labour time, although none of those interviewed knew how it was derived.) An overriding assumption of almost all discussion about automation is that productivity increases result from the substitution of machines for hourly production workers. That is, a reduction in factory jobs is *ipso facto* understood to mean a gain in productivity. Moreover, managers' effort to reduce the workforce is universally understood to reflect an interest in increasing productivity. A closer look coupled with a genuine concern for productivity would challenge this assumption and perhaps reveal other motivations at work. For, as Thomas Gunn argued in 1982, "Direct labor accounts for only ten to twenty-five percent of the total cost of manufacturing. . . . It is not clear that even a total replacement of blue collar workers by robots would [by itself] have much effect on the output of the factory or the cost of its products."[8]

John Simpson, Director of Manufacturing Engineering at the National Bureau of Standards, took this same message a bit further: "In metalworking manufacture, direct labor amounts to roughly 10 percent of total cost, as compared to materials at 55 percent and overhead another 35 percent. Yet, as of 1982, management was expending roughly 75 percent of managerial and engineering effort on labor costs reduction, as compared to 15 percent on materials cost reduction and 10 percent on overhead cost reduction. This is a striking disparity."[9]

It certainly is. As *Business Week* discovered in its 1982 survey of executives, few managers anticipated much use of the new equipment to displace management, even though such reduction in overhead, as Simpson suggests, would no doubt serve the goal of increased productivity.[10] But is it really the goal? The automatic causal link between innovation and productivity is ambiguous, and not only in terms of actual results but perhaps also in terms of intentions. Whenever managers are able to use automation to "fatten profits" and enhance their authority (by eliminating jobs and extorting concessions and obedience from the workers who remain) without at the same time increasing the social product, they appear more than ready to do so.

The next weak link in the causal chain connects productivity with competitiveness. The assumption is that greater productivity results in lower operating costs and thus lower prices and that this increases demand and enhances competitive strength. As we have seen, the increases in productivity due to automation are hard to assess with confidence. It appears also that when there have been productivity gains and, presumably, lower operating costs, these have rarely resulted in any reduction in prices. Wherever the gains are going, it is not to the consumer. But, for argument's sake, let us suppose that there have been lower costs and also lower prices. Does this guarantee greater competitiveness? Not really.

The truth of the matter is that competitiveness is tied less to operating costs and prices than it is to product quality, product design, marketing strategies, or service — as successful Japanese firms especially have demonstrated again and again. The ability to produce something more cheaply without these other factors might result only in the swelling of inventory, not sales. The simple assumption that lower costs through increased productivity result in more sales is naive and misleading. The Canadian study of banking automation, it should be emphasized, reported that managers acknowledged their "inability to use technology to achieve lasting competitive advantages." Instead, the investigators found that banks were introducing new technologies merely to appear to be competitive."[11] Here as elsewhere appearances can be deceiving.

Finally, to close the circle, there is the last assumption that increased competitiveness of firms results in prosperity for all. At the time Adam Smith formulated his ideas about how to increase the wealth of nations (late eighteenth century), he had good reason to assume some parallel between the prosperity of a company and the prosperity of its homeland, and the correlation continued to hold for a good long time.[12] But it no longer does. For today, most major manufacturing firms are multinational not only in terms of their market but also in terms of the scale of their productive operations. This global scale of operations is matched by a relatively unimpeded mobility. Firms have the ability to transfer production from one country to another, to close a plant in one place and reopen it elsewhere, to direct and redirect investment wherever the "climate" is most favourable. This mobility has resulted in a rupture between the health of a corporation and the prosperity of any one "host" nation (including its home base). That is, even when innovation and productivity do actually combine to increase the competitiveness of corporations, such competitiveness is no panacea, no guarantee of prosperity. Indeed, it has only better enabled the corporation to play one workforce off against another in pursuit of the cheapest and most compliant labour (which gives the misleading appearance of greater efficiency). Moreover, it has compelled regions and nations to compete with one another to try to attract investment by offering tax incentives, labour discipline, relaxed environmental and other regulations, and publicly subsidized infrastructure (such as roads and sewers). Thus has emerged the great paradox of our age, according to which those nations prosper most (attract corporate investment) by most readily lowering their standard of living (wages, benefits, quality of life, political freedom). The net result of this system of extortion is a universal lowering of conditions and expectations in the name of competitiveness and prosperity.

Not long ago I attended a conference with economists whose chief concern was competitiveness. Whenever anyone raised any other concern—health care, clean air and water, worker participation in decision-making—the economists harked back to their "bottom line," competition. Whatever the sacrifices, they contended in the hard-minded spirit of the "new realism," competitiveness had to be the top priority. I listened to this for perhaps too long and then finally asked these people if competition was the most important thing to them, and they said it was. I asked them if they would be willing to entertain any and all suggestions as to how we might enhance competitiveness, and they said they would. They were tough-minded, after all. So I suggested that they might consider reintroducing slavery.

Predictably, they balked at this suggestion and knew immediately that I was joking. After all, a civil war and centuries of violent struggle had succeeded in making slavery taboo in this country. Economists could no longer even raise a subject that had preoccupied their predecessors barely a century ago. Yet today, given the unprecedented mobility and power of multinational corporations and the all-enveloping ethos of extortion, the logic of competition is leading to similarly desperate conditions—shattered lives, deserted communities, abandoned hopes. And the economists are still allowed to discuss these "variables" in their equations of competitiveness—sunset industries versus sunrise industries; snowbelt versus sunbelt. Perhaps it will take another civil war to render these callous calculations taboo also, like slavery. Until then, however, enhanced competitiveness ought to give comfort to no one, except the investors.

Thus our circle of prosperity, upon closer inspection, appears less than compelling, and ideological imaginings alone cannot contradict the ambiguity of this promise. "The results," as the *Wall Street Journal* deftly understated it, "are mixed." But if the economic returns on automation remain ambiguous, the social consequences of automation do not. Unfortunately, they are all too clear.

PROGRESS FOR WHOM?

In their study of industrial automation my MIT colleagues found that just as there was little reliable data or certainty about the technical and economic viability of automation, so too there was "no commonly accepted calculus for assessing the societal costs and benefits of automation." They concluded, "The absence of such a calculus and of data to support it seriously weakens arguments in favor of automation."[13] But, unfortunately, neither data nor arguments have ever had much of a role in the drive to automate. Rarely challenged for either, the evangelists of automation have never had to marshal evidence or formulate arguments to defend their position; the power of their hegemonic ideology of progress has alone been sufficient to carry their campaign. The burden of proof, rather, has been borne alone by the critics of this campaign. Only the critics have had to come up with evidence and argument, usually just to have it dismissed without a hearing because it did not conform to the commonly held ideological presuppositions. Thus, it has been difficult for critics to provoke any serious — and much needed — debate.

The typical approach has been to speculate about the future, to estimate the number of jobs that will be lost or created. But such a crystal-ball approach is little more than a guessing game, and a biased one at that since the future looks grim, or rosy, depending upon who's looking and who's paying for the forecast. Every critical forecast is matched by an optimistic one. A more meaningful approach for estimating where we are headed is to examine historically where we've already been. Automation is not new. The term itself was coined in 1947 to refer to automatic transfer machinery in the auto industry, and the introduction of computer-automated equipment has been going on for some thirty years. The returns are in on this experience; what do they tell us, and what are their implications?

In addition to such a historical approach to this question, it is helpful also to be more specific about exactly who is the subject of our speculation. The impact of automation on society boils down ultimately to the impact of automation on particular people. There is no common calculus for assessing the societal costs and benefits of automation precisely because these costs and benefits are not borne and enjoyed by the same people, and one person's gain is another person's loss. Thus, in trying to assess the likely social consequences of this progress we must learn to ask: progress for whom?

In contrast to an economic analysis, the results of a historical and class analysis of automation are hardly ambiguous: by and large, the gains have gone to those with power, at the expense of those without it. There has been nothing automatic or inevitable about this outcome. It has followed not from any natural selection process or technical or economic logic, but, rather, from the political and cultural conditions that have prevailed (and continue to prevail). The differential impact of automation is readily seen in the history of the U.S. manufacturing industry over the last thirty years, judging from the official (and probably optimistic) data of the Bureau of Labor Statistics and the Department of Commerce.[14]

First, during this period, the value of capital stock (machinery) relative to labour doubled, reflecting the trend toward mechanization and automation. As a consequence, although the rate of productivity gain declined by half, the absolute output per person-hour increased by 115 per cent, more than double. But during this same period, real earnings for hourly workers, adjusted for inflation and cost of living, rose only by 84 per cent, less than double. Thus, after three decades of automation-based progress, workers are now earning less relative to their output than before. That is, they are producing more for less; working more for their bosses and less for themselves. (In addition to this relative decline in real wages, there has also been an absolute drop in those cases in which management has used the occasion of automation to shift workers from piecework to less lucratively measured day-work.)

Second, instead of seeing the number of working hours reduced in the wake of so-called labour-saving automation, workers have seen the average work-week either remain constant at forty hours or actually increase due to the compulsory overtime and shiftwork imposed by a management intent upon making the fullest utilization of its expensive new equipment (not to mention the additional hours resulting from the loss of paid holidays and sick-days extorted from unions by management with the threat of technological displacement). In short, labour-saving technologies have not been used to save workers' labour—meaning physical and mental effort—but rather to save capital labour—meaning workers (and wages). This double meaning of the word "labour" in a system in which one person can buy another person's labour power continues to confuse many observers.

Third, in the 1950s and 1960s, sociologists began to worry about the problem of "leisure": what would workers do with the time made available by the shorter work-week and less toil? But just as the shorter work-week never materialized, neither did the leisure — except in the form of enforced, involuntary leisure known as unemployment. Instead of being relieved of effort, workers have been relieved of their livelihoods. This human debris of progress was for a time obscured by the overall growth of a war-spurred expanding economy, and some of it was temporarily absorbed by the government-subsidized enlargement of the so-called service sector. But, once this boom began to bust, the outlines of all this progress became more visible. The proportion of employment in manufacturing relative to the total non-agricultural employment had already declined by a third, and it appeared that this was just the beginning of the end.

In 1978 the Organization for Economic Co-operation and Development (OECD) conducted a study of the impact of automation and concluded, "The evidence that we have is suggesting increasingly that the employment effects of automation, anticipated in the 1950's, are now beginning to arrive on a serious scale."[15] By 1982, as the economy continued to contract and the service sector itself came under an automation assault, even the traditionally sanguine General Accounting Office had to concede, "Whether automation will increase unemployment in the long run is not known." The Congressional Budget office was less circumspect. In 1983 that office estimated that by 1990, "A combination of automation and capacity cutbacks in basic industry will eliminate three million manufacturing jobs."[16]

This time a new escape hatch was invented; not the service sector but the "high-tech" industries, especially the manufacturers of automation equipment, would expand to absorb those displaced. But before too long, even executives in those "high-tech" industries had begun to express their doubts. "If you look at the long-term unemployment forecast and couple it to the whole issue of retraining, the problem is bigger than . . . we think," Peter Scott, executive vice-president of United Technologies Corporation, told the National Academy of Engineering in 1983. David T. Kearns, chief executive officer of Xerox Corporation, also acknowledged, "At least in the short term, I think there's a very real possibility that technology will put more people out of work before it puts them to work." Likewise *Iron Age* predicted, "There will be massive displacements." In 1983 *Business Week* finally concluded, "The number of new jobs created by high tech will fall disappointingly short of those lost in manufacturing," citing Nobel prize-winning economist Wassily Leontief's more colourful but no less austere conclusion that autoworkers had about as much chance to get jobs building robots as horses once did to get jobs building automobiles. "All the talk about new technology creating more and more manufacturing jobs, with workers using higher and higher skills, is a figment of the imagination," labour economist Sar A. Levitan also concluded in 1984: "There is just not that much there."[17]

Thus far, then, the consequences of automation for workers are no cause for optimism. The loss of income relative to output, the constant or expanding work-week, and the rising spectre of unemployment do not create a promising picture, as Leontief (one of the few economists with the courage to tell it as it is) has explained.

> [The] value of capital stock employed per man-hour in manufacturing industries in the U.S. . . . has almost doubled since the end of World War II. . . . Since the end of World War II, however, the work week has remained almost constant. . . . Concurrently, the U.S. economy has seen a chronic increase in unemployment from one oscillation of the business cycle to the next. The 2 percent accepted as the irreducible unemployment rate by proponents of full-employment legislation in 1945 became the 4 percent of New Frontier economic managers in the 1960's. The country's unemployment problem today exceeds 9 percent [1982]. . . . Americans might have [absorbed] potential technological unemployment by voluntarily shortening the work week if real wages had risen over the past 40 years faster than they actually have. . . . Sooner or later, and quite probably sooner, the increasingly mechanized society must face another problem: the problem of income distribution.[18]

Again, progress for whom? As Leontief suggests, the consequences have not been evenly distributed, have not been the same for everyone. For if the impact of automation on workers has not been ambiguous, neither has the impact on management and those it serves—labour's loss has been their gain. During the same first thirty-year period of our age of automation, corporate after-taxes profits have increased by 450 per cent, more than five times the increase in real earnings for workers. To the extent that there have been tangible benefits from automation, they have gone in only one direction: up. This fact was made painfully clear recently by the telling behaviour of the auto industry. In 1983, as the industry recovered from its temporary slump, General Motors paid six thousand of its executives almost $200 million in bonuses, averaging more than what an average GM worker makes in a year. Ford, not to be outdone, paid its top forty-five executives a half-million dollars each and its chairman $7.3 million. According to the *Los Angeles Times*, the record profits that made all this self-serving largesse possible resulted in part from the "introduction of modern equipment and sharp reductions in the automotive labor force."[19]

Once again, there was nothing automatic about these disparate outcomes. They followed from the social choices made by those who have had the power to choose. And, given this same constellation of forces, the future will more than likely be more of same: prosperity for the fortunate few and structural unemployment for the rest — precisely the starkly stratified have-have not society depicted by Kurt Vonnegut at the dawn of the automation age.

Calling Mr. Goodwrench

And still the snow-job-turned-sermon ideology of progress holds us under its stupefying spell, blinding us to this perilous prospect and automatically conjuring up a prettier picture. But there are signs that at least some people have begun to see through this mystifying haze and to recognize more clearly what is at stake. Early in 1984, the Louis Harris opinion survey research organization published the results of an extensive public poll they had conducted on the impact of technology on society. They discovered that people viewed this thing called progress differently depending upon where they sat.

> The difference between the public and the corporate executives on the matter of robots is a startling 54 percentage points. The tension between social classes is unmistakable. By 39 points, corporate executives are more optimistic about factory automation than are the people who work in factories. In addition, executives are more optimistic than skilled and unskilled labour as a whole by 41 points. These figures represent a potentially combustible mixture.[20]

Apparently, then, people are beginning to see automation madness for what it is and to recognize the management sermon on progress as the snow job it has always been. "If they have the right to say yes to technology and then move, we have the right to say no and prevent them from moving; that's equality," Frank Emspak, a local union leader at a large G.E. plant in Lynn, Massachusetts, declared.[21]

In other words, the progress of automation proceeds automatically at our expense only if, by our passivity, we allow it. Participation here demands defiance, defiance not only of the deceptive and disarming mythology of an automatic destiny but also of the destructive designs of those who peddle it. Such defiance alone, of course, is not sufficient. But without it we will never regain the confidence or the power to take this very serious matter of progress back into our own hands, where it belongs.

Notes

1. J.J. Jaeger, quoted in "The Automatic Factory," *Fortune*, October 1953, p.190.
2. For a thorough account of disinvestment, see Melman, *Profits Without Production*, and also Barry Bluestone and Bennett Harrison, *The Deindustrialization of America* (New York: Basic Books, 1982).
3. "Business Week/Harris Poll," *Business Week*, December 13, 1982.
4. "Preliminary Report on Industrial Automation," Center for Policy Alternatives, MIT, 1977.
5. "High Tech Track," *Wall Street Journal*, April 13, 1983; Thomas Gunn, quoted in "High Tech Track," *Wall Street Journal*, April 4, 1983.
6. "Costs of Technology Outstrip Benefits for Banks," *Computing Canada*, May 16, 1985 (courtesy Rene Jacques, Local 240, UAW Canada).
7. Henry Miley, quoted in Schlesinger, *Our Own Worst Enemy*.
8. Thomas Gunn, *Scientific American*, September 1982, p.129.
9. John A. Simpson, "The Factory of the Future," speech, Rensselaer Polytechnic Institute, Troy, N.Y., June 3, 1982.
10. Business Week/Harris Poll, *Business Week*, December 13, 1982.
11. "Costs of Technology Outstrip Benefits for Banks," *Computing Canada*.
12. Adam Smith, *The Wealth of Nations* (New York: E.P. Dutton and Co., 1933).
13. "Industrial Automation," final report, Center for Policy Alternatives, MIT, 1978.
14. The statistics on labour productivity, wages, hours, and employment are from the U.S. Department of Labor, Bureau of Labor Statistics; the data on corporate profits are from the U.S. Department of Commerce, Bureau of Economic Analysis.
15. OECD forecast, cited in *The New York Times*, July 5, 1978, p.D1. See also Norbert Wiener's 1949 Letter to Walter Reuther, in our Appendix VI.
16. GAO forecast, cited in *Washington Post*, July 27, 1982; Congressional Budget Office Forecast, cited in *Wall Street Journal*, April 3, 1983.
17. Peter Scott (United Technologies executive), quoted in *NAS/NAE News Report* (National Academy of Engineering), May/June 1983, p.23; David Kearns (Xerox CEO), quoted in *Rochester Review* (University of Rochester), Winter 1984; "Factory of the Future," *Iron Age*, February 25, 1983; *SQ Business Week*, March 28, 1983; Sar Levitan, quoted in *The New York Times*, March 25, 1984.

18 Wassily Leontief, "The Distribution of Work and Income," *Scientific American*, September 1982.
19 *Los Angeles Times*, May 2, 1984.
20 *The Road After 1984*, Louis Harris and Associates, Inc., 1983.
21 Frank Emspak, quoted in *Multinational Monitor*, March 1984, p.18.

THE HEARINGS ON INDUSTRIAL POLICY: A STATEMENT

CHAPTER 6

Good morning. I am David Noble, associate professor of the history of technology at MIT and currently on leave as curator of the Division of Mechanisms at the Smithsonian Institution, here in Washington.[1]

I have recently completed a study of the history of machine-tool automation in the United States in the postwar period. I should emphasize I am focusing on metal-cutting equipment and excluding metal-forming equipment. While I was doing this study I noticed that people generally attribute a greater degree of sober rationality to industry decision-makers than is actually the case. Like you and I, these people are good people — dedicated, hard-working and resourceful — but, also like you and I, they are sometimes moved by unseen drives, habits, enthusiasms, and, yes, fantasies.

I would like to suggest that as far as the machine-tool industry in the United States is concerned, these compulsions have perhaps got the best of them. Some of their compulsions are manifest already this morning. We all prefer to look elsewhere for the source of our problems, to Japanese policies or to something else; but more often than not, when we seriously confront the problem, it turns out to be us.

I don't mean to imply, I should emphasize at the outset, that this industry has been unique in this regard or that the collective compulsions I will describe account by themselves for all of this industry's

---

This chapter is a slightly revised version of a statement made to the Hearings on Industrial Policy before a Congressional Subcommittee of the 98th U.S. Congress.

problems. I don't want to slight the factors that have already been cited by other panelists.

One might suspect that the machine-tool industry is in the business of building and supplying metal-working machinery to manufacturers, which will enable them most efficiently to produce consumer goods. Certainly this must be considered a socially valuable enterprise, the guts of any machine-based industrial society.

Visitors to factories of the machine-tool and other metal-working industries would no doubt be impressed by the dedication, capability, and no-nonsense attitude of the people who work there. Nonetheless, I think these people have been caught up in certain collective compulsions, obscured in their day-to-day discourse and deliberations, their deals and deadlines, and these compulsions have been counterproductive to an otherwise socially beneficial enterprise.

What are these unexamined compulsions, compulsions that have influenced the industry beyond its genuine interest in manufacturing the finest possible machine-tools? First, of course, is the profit motive. We would all agree that these firms are interested in making money at least as much as they are in making machine-tools, and that they view the latter as the means to the former.

It is commonly assumed that in order to make a profit, firms in this industry must build the most economically competitive and technically viable machines, and that this is the surest route to prosperity. As I will try to show, this is not always the case, and when there is a choice between the best machines and the highest profit margins, the firms in this feast and famine industry have tended to opt for profits.

The second compulsion is what I would call the "machine mentality": the understandable but nevertheless self-serving belief that whatever the problem, a machine is the solution. This manifests itself in a preference for, and tireless promotion of, capital-intensive methods and in the widespread but mistaken belief that the more capital intensive the process of production, the higher the productivity.

This compulsion tends inescapably toward the elimination of skilled and unskilled labour alike and toward the reduction of direct labour costs, but it does not necessarily mean a reduction in unit cost. This compulsion also gives rise to fantasies of the automatic factory.

This compulsion is interwoven with the third major compulsion, which I call managerialism. This is the assumption that management control over production is the *sine qua non* of efficient production, and it leads typically to the single-minded pursuit of management control as an end in itself.

The machine-tool industry subscribes to this philosophy within its own shops and is reinforced in this orientation by the routine specifications of it customers—who are always managers in other industries. The machines it produces, therefore, are designed to maximize management control and minimize shop-floor intervention. Here, too, the tendency is toward the automatic factory or, more explicitly, the workerless factory.

Finally, the fourth compulsion is what can best be called enthusiasm. This I would describe as an enchantment, not with production or profit or even management control, but with the elegance and perfection intrinsic to the machine systems themselves. The best machines are thought to be those in which sublime control is vested in the designer, who has eliminated any and all chance for so-called human error.

This mentality, always present in the industry, was exaggerated immeasurably in the postwar period by the unprecedented entry of electronics, computers, and systems engineers. While this technical enthusiasm might well complement the drive toward the automatic, workerless factory, it does not always contribute to commercially viable production; it could even obstruct it. In short, experienced people in the machine-tool industry might seem to others to be archetypal businessmen, knowledgeable, and practical, with both computer ability and grease under the fingernails, but they too are caught up in rarely acknowledged compulsions.

The compulsions I have described are not new to the industry. But in the past they have tended to counterbalance each other, and to be tempered by the very real rigours of competitive production. Such production, moreover, has historically been rooted in a tradition of highly skilled and fiercely independent labour. Managers in the industry were heavily dependent upon an innovative, resourceful, and well-organized workforce, and, indeed, most managers and engineers had typically been recruited from this workforce. Thus the now-mythical practicality and well-roundedness of the decision-makers in this industry have a basis in fact.

What changed this, and magnified the collective compulsions to the point where they became literally counterproductive? In part it was the postwar boom, which, despite periodic slumps, was characterized by a relative lack of foreign competition. Remember that the major machine-tool competitors today were virtually destroyed during the war; it took decades for Japan and Germany, for example, to rebuild. During that time the U.S. machine-tool industry — the justly celebrated foundation of the "arsenal of democracy" — had a comparatively easy time of it. With the rigours of competition reduced, fantasies were let loose. Enter the U.S. Air Force.

If industrial policy in Japan has been administered by the Ministry of International Trade and Industry [MITI], industrial policy in the United States, in the absence of any formal policy, has been determined ad hoc but de facto by the Department of Defense.

With regard to production, the Department of Defense is driven by its own peculiarly military compulsions: performance, justified in the name of overriding strategic and security considerations whatever the cost or competitive consequences; and command, the imperatives of a hierarchical chain of authority in which all control is centred at the top and all actions below are specified in detail. Needless to say, the practical requirements of commercial production do not receive top priority in this rarefied realm, nor perhaps should they.

The problem is that the regimen and discipline characteristic of the military came increasingly to affect much of the machine-tool industry, as the military-industrial complex that President Eisenhower described so well steadily solidified in the postwar period. The machine-tool industry came under particularly heavy military influence, with the result that its own wayward compulsions were exacerbated by military predilections, and its own historic strengths and virtues were distorted beyond recognition.

The traditional quest for profits by means of economical production and commercially competitive quality machinery gave way to the rush for a quick and easy return through lucrative cost-plus contracts for oversophisticated dream machines. Commercial competence yielded to military extravagance, market responsiveness to Pentagon politics. And this trend was reinforced not only by military performance specifications but also by the military's undue reliance upon university-based technical expertise and the indulgent patterns of research and development that arose during the war.

In the machine-tool industry the result was an encouragement of unreasonable technical enthusiasm, and a shift away from the shop floor as a repository of innovative and practical ideas toward the laboratories of electronic systems and computer engineers and mathematicians, who knew little about the practical realities of machinery or production and cared less.

Finally, the military command imperative, and correlative fixation on closed-loop control, institutionalized management's penchant for centralized control and the fetish for machinery designed to facilitate it. While compatible perhaps with the compulsions of command and control, such machinery was no guarantee of either efficient production or commercial success.

In short, under the stimulus of the U.S. Air Force, which underwrote and oversaw nearly the entire development of computer-based machine-tool manufacturing, the U.S. machine-tool industry took flight, so to speak, lost its shop-floor basis and, with it, the practical common sense that had formerly been its hallmark and guarantor of prosperity.

Today these postwar delusions have come to seem completely natural. People working in the machine-tool industry have grown accustomed to the mentality and do not question it any more than fish wonder about water; they just swim in it and, I might add, drown in it. Many in the industry have known no other environment, and the recent demand for competitive, accessible, economic products has caught them unaware and woefully unprepared.

The people are not to blame, however. They were not born incompetent but were trained that way; theirs is a learned incompetence which, unfortunately, has served them well in their postwar artificial habitat. Thus, however well-intentioned, hard-working, and inventive, they are destined in this limited setting merely to go on diligently and dutifully mastering the mechanics, and compounding the crime.

This is not the place to go into great detail on these matters, except by way of illustration. . . . Consider the history of numerical control. Again, I want to emphasize I am talking about metal-cutting machinery exclusively.

There were several independent inventions along these lines in the 1940s, but nothing much happened until John Parsons came along. A manufacturing man from upper Michigan, Parsons conceived a relatively straightforward, practical device suitable for commercial production.

He managed to convince the U.S. Air Force that his idea was just what they needed to produce complex, close-tolerance components for their new high-performance aircraft then on the drawing board. Parsons received a development contract from the air force and went to MIT, among other subcontractors, for help with servomotors.

As it turned out, he got much more from MIT than he had bargained for. Within a year, the engineers there, none of whom had any experience with either metal-working or commercial production, had usurped his idea, forced him out of the project, and worked out a separate arrangement with the air force. . . . Before long, Parsons's dream of a commercially viable automated machine-tool gave way to a much more expansive concept, one even more compatible with MIT's interest in state-of-the-art computer control and the air force's fantasies of omnipotence.

Military extravagance combined with indulgent technical enthusiasms to yield a technology that was oversophisticated for the needs of all but a tiny fraction of the metal-working industry. It was also technically unreliable and prohibitively expensive for all but the military-subsidized aerospace industry.

The same thing happened in the case of numerical-control software. The air force, with its emphasis upon universal applicability and fancy five-axis control, insisted upon the development and exclusive use of the so-called APT system — also created at MIT — which effectively foreclosed for a decade or more the development of less cumbersome and more accessible numerical-control software systems.

Predictably, there was no warm reception for this air force/MIT technology once it hit the metal-working marketplace. No one would touch it. So the air force, still together with MIT, entered the promotion business. But this effort proved futile.

Finally, the air force decided to circumvent the existing market altogether and create a market that would be all its own. At taxpayer expense, it purchased and underwrote the costs of development, software, installation, and training for every first-generation numerical-control machine. At this point the machine-tool industry, which had with some exceptions previously remained aloof, entered the numerical-control business with a rush, since customers and profits were now guaranteed by air force largesse.

Although the air force was instrumental in originating numerical control, the patterns of development and use it fostered proved ultimately counterproductive. Thirty years after the unveiling of the first numerical-control milling machine at MIT in 1952, less than 3 per cent of the machine-tools in the United States were numerically controlled, as compared to early 1950 projections of 80 per cent for that same period.

As the machine-tool builders scrambled for military contracts, their criteria as to what constituted the best machinery had come to resemble narrow military criteria rather than what would work and what would sell in the metal-working industry as a whole. Thus, they tended to pay less attention to such things as competitiveness, cost, accessibility, and simplicity, until it was too late.

Meanwhile, of course, Japanese and German manufacturers, free of military blinders, began to turn out commercially viable equipment that was simple, accessible, reliable, and less expensive, along with the complementary software. That is why in so many shops in the United States today, while American flags fly outside, foreign machines are at work inside.[2]

A second illustration involves the abandonment not only of more accessible versions of numerical control but also of alternatives to numerical control itself. As we have seen, the profit orientation of the machine-tool industry led it astray once the highest profit rates were no longer immediately consonant with the production of commercially viable products. A similar distortion was generated by the military magnification of the managerial predilection for centralized, authoritarian control over production, a tendency reinforced also by the preferences of systems designers. Machinery was designed with the expectation that it would facilitate the removal of as much control from the shop floor as possible, thereby enhancing management authority over the production process and eliminating seemingly unnecessary worker intervention.

Unfortunately, the realities of production defy such an authoritarian approach to production efficiency, as evidenced by almost every case cited in the literature of industrial sociology. Moreover, machines designed with this approach in mind were of little use to the typical machinist-managed job shops, which constitute some 80 per cent of the metal-working industry. Foreign competitors, like Japan and Germany, have placed much more emphasis upon accessibility and shop-floor control and thus have more easily penetrated the lucrative U.S. job-shop market.

The emphasis upon management control among U.S. manufacturers led to the abandonment of a promising alternative to numerical control, one that might have gained them a much earlier foothold in the job-shop market for automated equipment: record-playback.

This concept, developed in the early 1940s, was consistently dismissed despite its technical and economic benefits, in large part because it would have left control over programming and production in the hands of machinists and unions. After 1950 it was almost entirely overshadowed by air force-sponsored numerical control, a much more complicated and expensive approach.

While it is true that record-playback methods were not suited for some of the more demanding military specifications — largely restricted to the manufacture of aircraft, missiles, submarines, and, today, B-1 bombers — there is no compelling technical or economic reason why this alternative approach could not have proved a great step forward for the run-of-the-mill job shop, which was for decades effectively barred from participation in the so-called numerical-control revolution.

I might add parenthetically that various updated — digitized — versions of record-playback control have recently begun to appear. For example, a practical-minded and shop-oriented colleague of mine at MIT, David Gossard, developed a system that substantially reduces programming and training time required for automated machining and renders it completely accessible to shop-trained machinists, but he could find no U.S. manufacturers who were willing to fly with it. Today, thanks to the resourcefulness of one of his Japanese students, his system is in use in Japan.

Finally, the authoritarian approach characteristic of the U.S. machine-tool industry management might well be creating an irreversible trend that will steadily deplete this nation's accumulated store of irreplaceable skills. In addition to producing machines to try to reduce the skill requirements of workers, the industry is at work also trying to produce the less-skilled people to match their automated machinery.

While NMTBA President James Gray warns hyperbolically that the "U.S. faces one of the greatest skill shortages in the history of the country," the NMTBA's own John Mandl, in a 1981 report to the air force, proposed that machinist apprenticeship time be cut almost in half in order to "upgrade the content in a narrower scope and reduce the skill levels required to operate and maintain automatic machine-tools."

I am sceptical about the claim that we now face any such skill shortage, but if the present trend continues, this could prove to be a self-fulfilling prophesy. I refer you to Neal Rosenthal's study on this.[3]

Having eroded the commercial competence of those in the machine-tool industry, our collective compulsions might some day rob us as well of the skilled, and still competent, workforce upon whom the U.S. machine-tool industry has always depended, and still depends.

What, then, is to be done? What can the public sector do in the face of these challenges? Do we protect the machine-tool industry from foreign competition? Help U.S. builders compete better by subsidizing a modernization of their industry? Let them die a natural death and make room for seemingly more competitive sunrise industries? I will look briefly at each of these suggestions in turn.

Houdaille Industries has called for sanctions against U.S. machine-tool users who purchase foreign computer-controlled equipment, while the NMTBA has called for protectionist measures in the name of preserving the domestic industrial base and, thus, national security.

I have no quarrel with protectionist measures so long as they mean the protection of people; that is, of society against the ravages of the morally blind market. But I am not convinced any longer that the protection of industries or firms necessarily means the protection of people, given the unprecedented mobility of capital and the accelerating drive toward automation. Protection for this industry does not automatically mean protection of jobs and revenues as it used to, especially given the dominance of the collective compulsions I have described.

Similarly, while a viable domestic industry is essential not only for national security but also for the industrial health of the nation, I am not convinced that protection from foreign competition would in itself guarantee such viability, again given the compulsions I have outlined.[4]

If protectionist measures were adopted in order to buy time and were coupled with a thoroughgoing reassessment of the systemic problems of the industry, something of lasting value would be gained. Otherwise, protecting the machine-tool industry would be a Sisyphean undertaking, with no end in sight.

Should the machine-tool industry be subsidized and modernized to achieve belated competition-worthiness? It has long been claimed that the firms in this industry are simply too small to underwrite their own research and development, their own improvement, and that therefore some outside support is required. This, of course, is precisely the role the air force has played since the war, a role that was played by the navy earlier in the century and by the army in the nineteenth century. Indeed, the air force is still playing this role with its ICAM and related programs.

All of these efforts are profoundly marked by the collective compulsions, and they give innovation a bad name. I am convinced that any further movement in this same direction, whether subsidized by military or civilian programs — and I would include the H.R. 5540 effort to revitalize the military-industrial-academic base in this category — would only make things worse. Again, without a thoroughgoing reassessment of criteria and objectives, of past habits and futuristic fantasies, any supposed modernization would only further cripple an already hobbling industry.

We must begin to ask ourselves: is an internationally competitive industry a good industry? Even if certain measures do enhance the international competitiveness of these firms, would this necessarily serve our domestic need for quality machinery? Progressivist critics have assailed this industry for its alleged "backwardness," deriding its "ma and pa" scale, its decentralized nature, its lagging labour-intensive methods, and they have argued that to compete effectively in the international market, the industry will have to become more concentrated and streamlined.

I am not convinced. I find nothing inherently wrong with ma and pa decentralized, small-scale, labour-intensive industries if they serve an important function. Indeed, it has been argued persuasively by Lewis Mumford and others that these traditional features of industry have been essential for the versatility, diversity, durability, and resiliency of not only U.S. industry but of industry throughout the world.

As a native Floridian, I well recall the dependence of NASA moonshot engineers on the resourcefulness and ingenuity of local job shops. If greater international competitiveness means an industry that is less attuned to the variegated and vital needs of the domestic manufacturing market, then it would only compound the problems already created by the collective compulsions.

Given the mobility of capital and the trend toward more highly automated production, there is no longer any guarantee that the competitiveness of firms will mean social prosperity domestically — what used to be called the "wealth of nations." For working people especially, who desperately require jobs and stable communities, competitive companies are not the reassuring panacea they once were. I would caution, therefore, against a preoccupation with international competitiveness. Indeed, I would suggest that international competitiveness is becoming another one of our collective compulsions. It should be seriously re-examined.

Finally, some observers have suggested that the domestic machine-tool industry ought to be allowed to die a natural death and that we should henceforth rely exclusively upon imports to meet our needs. This suggestion represents total surrender to the supposed wisdom of the principle of comparative advantage and to the belief, belied by reality, in the beneficence of the self-regulating market.

The fact of the matter is that there is no deliverance in free trade any more than there is salvation in high technology. Years ago, Kurt Vonnegut wrote that those who live by electronics die by electronics; the same could be said of those who live by the international market. The touted beneficent circle of prosperity, which links investment to innovation, innovation to productivity, productivity to competitiveness, and competitiveness to social prosperity, is now ambiguous at each link. There is no longer any guarantee of social welfare along this route, given the mobility of capital and the global reach of multinational corporations.

What remains unambiguous, however, are the social consequences of continued conformity to our collective compulsions: structural unemployment, social dislocation, tightening controls over people at work, and the erosion of traditional skills and thus also of quality production, industrial capability, and commercial competence.

Rather than focus our attention upon firms or industries, therefore, and rely upon faith that their fortunes will translate somehow into social prosperity and human welfare, we ought to consider what we want our policies to achieve: full employment, stable communities, a sound infrastructure, regional integrity, decent health, education, food, clothing, and housing for all citizens, and fuller, not more restricted, democracy.

These, it seems to me, must be the constants in any policy equation — not the variables that they are now, when the constants are profit maximization, military expansion, and technological fetishism. Industrial policy really is social policy, and the indicators of success should not be the degree of industrial competitiveness or the number of machines or the return on investment, but rather the level of satisfaction of our citizens.

A good look at this level today would give thoughtful people pause and cause them to reflect upon their firmly held but heretofore unexamined convictions. I believe that once our social needs are met, our industries will again become sound. But to achieve both goals, we must demilitarize our economy, democratize our industry, and reduce the play of our underlying compulsions.

NOTES

1 See *Industrial Policy: Hearings before the Subcommittee on Economic Stabilization of the Committee on Banking, Finance and Urban Affairs*, House of Representatives, First Session, Part 2, July 13, 14, 18, 26, 27 and 28, 1983, Serial No. 98-44; printed for the use of the Committee, U.S. Government Printing Office, Washington, D.C., 1983, pp.566-75.
2 Mr. Noble elaborated in his corrected transcript: U.S. numerical-control equipment manufacturers' difficulties in penetrating the significant small- and medium-sized civilian job-shop market clearly predated the challenge of Japanese competition, a fact well illustrated in the hearings on numerical control, before the Subcommittee on Science and Technology of the Select Committee on Small Business, U.S. Senate, 92nd Congress, 1st session, June 24, 1971.
3 Neal H. Rosenthal, "Shortages of Machinists: An Evaluation of the Information," *Monthly Labor Review*, July 1982.
4 For a glimpse of the negative consequences of these compulsions within the military arsenals themselves, see, for example, the General Accounting Office (GAO) reports on "Use of Numerically Controlled Equipment," June 26, 1975, and "Manufacturing Technology," June 3, 1976. They point out that these compulsions have produced severe shortcomings in the military as well as civilian industry.

CHAPTER 7
THE RELIGION OF TECHNOLOGY: THE MYTH OF A MASCULINE MILLENNIUM

In their homes, much glass and steel. Their cars
are fast—walking's for children, except in rooms.
When they take longer trips, they think with contempt
of the jet's archaic slowness. Monastic
in dedication to work, they apply honed skills, impatient of
less than perfection. They sleep by day
when the bustle of lives might disturb their research,
and labor beneath fluorescent light in controlled environments
fitting their needs, as the dialects
in which they converse, with each other or with
the machines (which are not called machines)
are controlled and fitting. The air they breathe
is conditioned. Coffee and coke keep them alert.
But no one can say they don't dream,
that they have no vision. Their vision
consumes them, they think all the time
of the city in space, they long for the permanent colony,
not just a lab up there, the whole works, malls, raquet courts,
hot tubs, state-of-the-art
ski machines, entertainment . . . Imagine it, they think, way
out there, outside of 'nature,' unhampered,
a place contrived by man, supreme
triumph of reason. They know it will happen.
They do not love the earth.
— *Those Who Want Out*, Denise Levertov, 1984

In a number of short essays on technological progress, Christopher Lasch pointed out that there might be something profoundly irrational at the core of what is assumed to be *the* quintessentially rational enterprise. "The intellectual basis . . . of the modern cult of technology," Lasch noted, is "the celebration of disembodied intelligence," an "incorrigibly escapist . . . fantasy of total control, absolute transcendence of the limits imposed on mankind." By implication, Lasch was suggesting that the technological endeavour upon which we have come to rely for the preservation and enlargement of life betrays an impatience and even disdain for life, a contempt and defiance of our bodily, that is mortal, earthly existence. The seeming practical pursuit of utility is in reality a quest for transcendence. Our worldly ends, oddly (and dangerously) enough, have become dependent upon what are, at bottom, other-worldly means.

Other writers have made similar observations, describing this technological mentality as an essentially religious phenomenon. Lewis Mumford long ago linked our modern malaise to Western culture's enduring "faith in the religion of the machine," while, more recently, Wendell Berry despaired over "our curious religious faith in technological progress," despite the destructive consequences. This chapter briefly explores the historical evolution of this religion of technology, which has so taken hold of our imagination and, hence, our future.

This religious spirit of modern technology is apparent today in the hallmark enterprises of the age, from the enchanted exploration (and alteration) of the heavens to the obsessive preoccupation with "a new creation," the improved product of genetic manipulation and artificial means. It is no mere coincidence that the father of the U.S. space program, Wernher von Braun, was himself a born-again Christian, or that NASA has long been a haven for Christian evangelicalism, or that all the moon astronauts were devout Christians. Nor is it unusual to find the enthusiasts of genetic engineering and artificial reproduction, in avid pursuit of the "holy grail" of the human genome and the artificial womb, exulting in their god-likeness. (The director of the Human Genome project, Francis Collins, is another born-again Christian.) The transcendent trajectory is obvious in the discourse of the artificial intelligentsia, vanguard of automation and robotics, who speak longingly of disembodied existence, of "mind transfers," "downloading minds into machines," machine-based "immortality," "artificial life," and "postbiological evolution." A recent history of artificial intelligence (AI), by practitioner Daniel Crevier, contains a chapter on "AI and Religion" in which the author insists that AI is not incompatible with the Christian notion of resurrection of the soul.

The escapist impulse and fantasy of omnipotence that characterize the religion of technology are today perhaps most explicit among the apostles of cyberspace, the designers and evangelists of the "information revolution." As computer-industry consultant Michael Heim acknowledges, "Our fascination with computers . . . is more deeply spiritual than utilitarian." That fascination reflects a striving to emulate the "all-at-once-ness of divine knowledge." According to Heim, "What better way, then, to emulate God's knowledge than to generate a virtual world constituted by bits of information. Over such a cyber world human beings could enjoy a god-like instant access. . . . When on-line, we break free . . . from bodily existence." The designers of one of the first computer networks, a community bulletin board developed in 1978 for the San Francisco Bay area, opened their prospectus with the words: "We are as gods and might as well get good at it." As one sociologist described their efforts, they were charged with a "technospiritual bumptiousness, full of the redemptive power of technology."

Architect Michael Benedikt, president of Mental Tech, a software design company, and editor of a major anthology on cyberspace, has argued that cyberspace is the electronic equivalent of the imagined spiritual realms of religion; what people have heretofore sought in religion they will find in cyberspace. Religions, Benedikt writes, are fuelled by "the resentment we feel for our bodies' cloddishness, limitations, and final treachery, their mortality. Reality is death. If only we could, we would wander the earth and never leave home; we would enjoy triumphs without risks and eat of the Tree and not be punished, consort daily with angels, enter heaven now and not die." Cyberspace, according to Benedikt, is the dimension where "floats the image of the Heavenly City, the New Jerusalem of the Book of Revelation. Like a bejeweled, weightless palace it comes out of heaven itself . . . a place where we might re-enter God's graces . . . laid out like a beautiful equation."

At the University of Washington's Human Interface Technology Laboratory, among the vanguard of virtual reality, the vision explodes into ecstasy. "Cyberspace can be seen as the new bomb," exclaims one Lab member, "a pacific blaze that will project the imprint of our disembodied selves on the walls of eternity.... On the other side of our data gloves, we become creatures of colored light in motion, pulsing with golden particles.... We will all become angels, and for eternity! ... Cyberspace will feel like Paradise." It will be "a space for collective restoration," in which we will rediscover the "habit of perfection."

Where did this religious spirit of modern technology come from? How did the so-called "useful arts," the most material of activities, get tied up with the spiritual? How did the practical preoccupation with nature become also, or rather, a supernatural pursuit? To answer such questions is also to explain how the useful arts came to be exalted, indeed worshipped, in our culture, after millennia in which such humble activities were culturally disdained because of their association with manual labour, slaves, and the work of women. It is also to explain how technology came to be identified so exclusively with men, evocative of an almost primordial masculinity, when throughout most of human history such practical affairs were identified at least as much with women as with men.

At the heart of this change was a major ideological transformation, which began among medieval monks, whereby the useful arts became belatedly implicated in the Christian mythology of redemption. Once invested with such spiritual significance, which changed their purpose from survival to salvation, the heretofore lowly arts were elevated in status to become worthy vehicles of elite male identity and devotion, with dramatic consequences. The emergence of Western technology as a historic force and the emergence of the religion of technology were thus one and the same phenomenon, as they have remained ever since.

The Christian myth of redemption, a variant of an ancient and universal theme, entails a return to origins and a recovery of lost perfection. In particular, it involves the restoration of the "image-likeness of man to God" as described in the Book of Genesis. According to the Judao-Christian story, Adam was created in God's image and, accordingly, had a share in divinity; Adam in Eden was thus given dominion over all other creatures, full knowledge of the natural universe, and even a role in creation (symbolized in his naming of the beasts and in his "giving birth" to, and also naming, Eve). His existence was self-sufficient and eternal. Alas, in the Fall Adam lost this divine likeness, and with it his true knowledge, self-sufficiency, dominion, immortality, and role in creation. For his sin of aspiring to be as God, rather than just in God's image, he was now cursed with the punishments of labour and death. The Christian myth of redemption involves a recovery, by various means, of Adam's lost perfection and, with it, his lost position and powers.

The Christian story of the creation, Fall, and restoration of Adam is a decidedly masculine one, no doubt reflecting the privileged bias of its male authors. God is the Father, who creates a son in His image, and it is this masculine divine image that is lost and recovered. In the first book of Genesis there is ambiguity on this point: "So God created man in his own image, in the image of God created he him; male and female created he them" (Gen. 1:27). But however much heterodox commentators used this passage to assert a positive female role in the story, orthodox commentary either ignored it, focusing rather on the preceding passage, "Let us make man in our image, after our likeness" (Gen. 1:26), treating it as allegory (for Augustine, male means Christ; female means the Church) or emphasizing instead the (for that reason more familiar) creation story of the second chapter of Genesis, in which God first created Adam, and then Eve, from Adam.

According to orthodox interpretation, then, which became the dominant Western myth, Eve was not created in God's image and thus neither had nor lost such perfection; nor therefore could she recover it. Woman was left out of the core Christian endeavour. This was made explicit by St. Paul; in his first letter to the Corinthians he insists that women who pray or prophesy must cover their heads, while men who do likewise must not. "For a man indeed ought not to cover his head, forasmuch as he is the image and glory of God: but the woman is the glory of the man" (1 Cor. 11:7). For Paul, moreover, Christ, the "second Adam," was the image of God incarnate; knowledge of Christ was thus, in effect, a kind of self-knowledge, a reminder of the original state. Christ was male. If woman was excluded from original divine likeness, so, by definition, must she be excluded from its recovery. The Book of Revelation, the guidebook for two thousand years of such expectation, explicitly restricts redemption in the Millennium to men, to "they which were not defiled with women; for they are virgins" (Rev. 14:4).

Of course, woman did get a role in the Christian myth of redemption, but only in a negative sense. According to Genesis 2, it was Eve who caused the Fall in the first place—Adam himself blamed her for the transgression. Thus, because of this act by woman, who had no divine likeness to lose, men lost their exalted role in Creation, and with it their knowledge, dominion, and immortality. By implication, women have ever since been perceived as the perpetual impediment to its recovery. The Christian promise of redemption, then, is for men only; the pursuit of perfection is a male effort to return to a male-only world. The masculine Millennium is the recovery not simply of Paradise, but of Eden before Eve.

It is thus not at all surprising that, like today's enthusiasts of space exploration, artificial intelligence, cyberspace, artificial life, artificial reproduction, and genetic engineering, most of those who devoted their lives to the restoration of perfection were men. Moreover, they were overwhelmingly men who inhabited worlds without women—monks and celibate clerks who experienced in their everyday lives a foretaste of paradise. (For a history of the evolution of this womanless world, see my book *A World Without Women: The Christian Clerical Culture of Western Science*.) At first, among the church Fathers and founders of monasticism, the pursuit of perfection entailed pious devotion, asceticism, spiritual contemplation, brotherhood, and study. The Benedictines later added manual labour to the list. In the so-called Carolingian renaissance of the ninth century, a period of significant development in the useful arts, these too came to be included among the vehicles of salvation. Thus there emerged, from a world without women and in a spirit of transcendence, the project of perfection we now call technology.

The earliest known instance of this new view of the arts is in the work of the foremost Carolingian scholar, Johannes Scotus Erigena. His was the first known use of the term "mechanic arts," which he used generically to define as a distinct category of human activity including all of the various crafts—a forerunner of the terms "useful arts" and "technology." Second, he departed from the ancient and Augustinian traditions by dignifying these heretofore lowly endeavours with their inclusion in his classification of knowledge. Third, he explicitly assigned them to men, as opposed to women. And, finally, he "Christianized" such activities by investing them with spiritual meaning; he identified them as vehicles of redemption, and the learning and practice of them as acts of recovery. "All men by nature possess natural arts, but because, on account of the punishment for the sin of the first man, they are obscured in the souls of men. . . . In teaching we do nothing but recall to our present understanding the same arts which are stored in our memory." The practical arts, in Erigena's view, constituted "man's link with the Divine, their cultivation a means to salvation." Erigena assumed that such acts of recovery were for men only and that restored perfection was masculine: "At the Resurrection," he proclaimed, "sex will be abolished and nature made one. There will then be only man, as if he had never sinned."

This ideological transformation of the useful arts proved enormously influential, especially within all-male monastic orders, in which, in the absence of women, elite men came to assume the burdens of production. By the twelfth century these men had created a veritable industrial revolution, substituting waterpower for womanpower. Inspired by Erigena, the Augustinian monk Hugh of St. Victor elevated the status of craft knowledge by including it in his own influential classification of learning. He likewise emphasized the new spiritual significance of the arts, the restoration of powers and perfection lost in the Fall. "This, then, is what the arts are concerned with, this is what they intend, namely, to restore within us the divine likeness."

The monastic elevation and masculinization of the useful arts as a medium of redemption were coupled in the high middle ages with a revived and revised millenarianism, based upon Biblical prophecy and, especially, the Book of Revelation. This last book of the Bible, which foretells of a thousand-year reign of the returned Messiah and an elite corps of redeemed saints (the Millennium), is, in effect, the happy ending of the first, Genesis, in that it promises a recovery of man's godliness. In the first twelve hundred years of Christianity, this revelation kindled passive hopes of redemption through divine intervention and the return of the Messiah. Now, in a revised interpretation formulated by the Calabrian Cistercian abbot Joachim of Fiore, which was to become the most influential prophetic system in the West until Marxism, millenarian expectation became active rather than passive. Joachim read the Book of Revelation as a chronicle of, and practical guide to, history, in which mortals played a vital role in preparing for and thus bringing about the Millennium.

Joachim described history as unfolding in three successive stages, the third being one of transition to perfection. The agents of this transition were the spiritual men, whose lives were devoted to millenarian preparation; these were the vanguard of humanity, bridging the chasm between the fallen and the redeemed. Joachim assumed, of course, that his *viri spirituales* would be males, and celibate, given the rigorously ascetic monastic world that he inhabited and the prescription of the Book of Revelation. For him, the monks, in particular his fellow Cistercians, were the new spiritual men who would bring about the Millennium by way of contemplation and spiritual illumination. Almost immediately after his death, however, the mantle of the third age was claimed by another breed of spiritual men, the mendicant friars. The Franciscans, especially the more radical or "spiritual" Franciscans, sought to bring about the Millennium not just through cloistered contemplation but as evangelical missionaries in the world. In addition, they added another dimension to such millenarian preparation, the advancement of the arts and sciences.

In this, the first and foremost proponent was the Franciscan Roger Bacon, celibate inhabitant of the womanless worlds of the mendicants and the medieval universities. Steeped in Biblical prophecy and especially Joachimite millenarianism, Bacon strongly advocated the development of the arts for the explicit purpose of preparing for the Millennium by recovering man's original knowledge. "All wise men believe that we are not far removed from the times of Antichrist," Bacon warned the Pope, and he counselled that this apocalyptic challenge might best be met "if prelates and princes promoted study and investigated the secrets of nature and of art." Bacon believed that the arts, the birthright of the "Sons of Adam," had once been fully known by men ("the saints at the beginning"), that they had been lost to sin but had already been partially regained, and that they might yet be fully restored through sustained and righteous effort. Like Erigena and Hugh of St. Victor, Bacon promoted the recovery of the arts as an aspect of the recovery of original perfection.

Other Franciscans followed Joachim's and Bacon's lead, among them the fervently millenarian triumvirate of Catalan science, Ramon Lull, Arnald of Villanova, and John of Rupecissa. The mendicant missionaries, moreover, influenced another breed of spiritual men, the great explorers of the age of discovery, the archetypes of human striving so emulated by today's astronauts and artificers, who claimed for themselves the sanction of prophecy. Their expeditions, which expressly excluded women, epitomized at one and the same time both the advance of the useful arts — in shipbuilding, navigation, metallurgy — and the active quest for the earthly paradise.

Here the central figure was of course Columbus, intrepid discoverer of the New World and master of the mariner's arts. Columbus was profoundly influenced by the monks and friars: he patronized monasteries, prepared for his voyages within their cloisters, wore the habit of a Minorite friar, became a Franciscan tertiary, and was buried in a Carthusian monastery. He was also, like many of his Spanish contemporaries, a devout millenarian; he immersed himself in the Book of Revelation and Joachimite commentaries, wrote his own Book of Prophecies, and, most important, viewed his own endeavours, which he called "the enterprise of Jerusalem," as the fulfilment of prophecy. "God made me the messenger of the new heaven and the new earth," Columbus wrote, "of which he spoke in the Apocalypse of St. John (Book of Revelation) after having spoken of it through the mouth of Isaiah, and he showed me the spot where to find it." Columbus was convinced that he had indeed recovered the "terrestrial paradise" and symbolized this restored dominion by compulsively naming, as Adam had, all that he surveyed.

The discovery of the New World fostered both the advancement of the useful arts and the masculine millenarian dreams with which those arts were now bound. In the fifteenth and sixteenth centuries, Renaissance humanists, magi, and illuminati viewed themselves as agents of religious revival and sought in ancient and pagan lore and the advancement of the arts the means of both purification and restoration. Thus Pico della Mirandola, a Joachimite supporter of Savonarola, laboured to recover the lost secrets of hermetic natural philosophy and the occult arts, and both Agrippa and Paracelsus drew inspiration from Joachimite commentary.

Agrippa argued that the original power over nature that Adam lost could be regained by the purified soul, the magus. "Once the soul has attained illumination, it returns to something like the condition of Adam before the Fall, when the seal of God was upon it and all creatures feared and revered man." Paracelsus, the founder of pharmacology, wrote, "When the end of the world draws near, all things will be revealed.... Blessed be those men whose reason will reveal itself.... For the light of nature is in us and this light is God. Our mortal bodies are vehicles of the divine wisdom.... Therefore, study without respite, that the art may become perfect in us." The great artist Albrecht Dürer, a contemporary of Paracelsus, shared both his advocacy of the arts and his millenarian enthusiasm. Inspired by the early rumblings of the Reformation, Dürer's first great work was his Revelation of St. John, a vivid depiction of the masculine Millennium.

The Reformation rekindled millenarian hopes as never before, as much among the respectable elite as among the downtrodden poor. In the eyes of many reformers the rupture in the Church signalled the coming apocalypse, and the advent of the Millennium. And with the renewed promise of recovery from the Fall and the restoration of Adam's rightful dominion came an intensification of misogynist sentiment, with a focus upon the "mother of our miseries," Eve. "Oh! Why did God, the Creator wise, that people'd highest Heaven with Spirits masculine, create at last this Noveltie on Earth, this fair defect of Nature, and not fill the World at once with men as Angels without Feminine, or find some other way to generate Mankind?" John Milton wrote in "Paradise Lost." There are no women in his "Paradise Regained."

The Reformation also fuelled as never before the advance of the arts in the pursuit of salvation. As Ernst Benz observed, "The modern technological revolution ... converted the Christian expectation of the coming of the Kingdom of God into a technological utopia." In the blessed utopias of Thomas More and the Joachimite millenarians Tomaso Campanella and Johannes Andreae, the practice of the arts was coupled with fraternal community in the pursuit of salvation. In Andreae's "Christianopolis," for example, the arts were encouraged "not always because necessity demands it, but ... in order that the human soul might have some means by which ... the little spark of divinity remaining in us may shine brightly." The arts, Andreae maintained, allowed men "to return to themselves." The Rosicrucian manifestos, which inspired a vigorous scientific and technological reform movement throughout Europe, proclaimed, "God has revealed to us in these latter days a more perfect knowledge, both of his Son ... and of Nature. He has raised men imbued with great wisdom who might renew all arts and reduce them all to perfection, so that man might understand his own nobleness and worth. ... God hath certainly and most assuredly concluded to send a grant to the world before her end, which presently shall ensue, such a truth, light, life and glory as the first man Adam had, which he lost in Paradise." Predictably, women were excluded from this benighted brotherhood.

The Rosicrucian message was heeded by generations of ardent and influential utopian reformers. By far the most influential among them was Francis Bacon. More persuasively than anyone before or since, he defined the masculine millennial project of modern technology, proclaiming that the development and elevation of the arts formed the key to the advancement of human knowledge, which itself was both the surest sign of, and the best preparation for, the restoration of perfection — "the entrance into the kingdom of man, founded on the sciences, being not much other than the entrance into the kingdom of heaven." Bacon insisted, "It was not that pure and unspotted natural knowledge whereby Adam gave names to things agreeable to their natures, which caused the Fall," but rather the quest for moral knowledge. He believed the recovery of Adamic natural knowledge was destined and (citing the prophecy of Daniel) foretold, and that it signalled what he described as "the great restoration of the power of man over the universe," wherein the "commerce between the mind of man and the nature of things . . . might be restored to its perfect and original condition." Bacon's overriding aim, according to historian Paolo Rossi, "was to redeem man from original sin and reinstate him in his prelapsarian power over all created things."

"It is not the pleasure of curiosity," Bacon wrote, "nor the quiet of resolution, nor the raising of the spirit, nor victory of wit, nor faculty of speech, nor lucre of profession, nor ambition of honor or fame, nor enablement for business, that are the true ends of knowledge . . . but it is a restitution and reinvesting (in great part) of man to the sovereignty and power (for whensoever he shall be able to call creatures by their true names he shall again command them) which he had in his first state of creation. . . . For man by the Fall fell at the same time from his state of innocence and from his dominion over created things. Both these losses can even in this life be partially repaired, the former by religion and faith, the latter by arts and sciences." In keeping with the dominant masculine mythology of Christian redemption, he entitled an early draft of "The Great Instauration," his magnum opus, "The Masculine Birth of Time." Addressed in an avuncular style to "my son," he bequeaths "my only earthly wish, namely to stretch the deplorable narrow limits of man's dominion over the universe to their promised bounds," thereby to create "a blessed race of Heroes or Supermen." He writes, "Take heart, then, my son, and give yourself to me so that I may restore you to yourself."

The bold Baconian project, and with it the religion of technology, was carried forth by a host of likeminded visionaries, among them the circles of the millenarian reformers John Comenius and Samuel Hartlib. Many of the virtuosi who gave rise to the Royal Society followed in their footsteps. Robert Boyle, father of experimental science and chemistry, wrote expectantly about the enhancement of knowledge that would reward those redeemed in the Millennium, "the great renovation of the world": "It is likely that as our faculties will, in the future blessed state, be enlarged and heightened, so will our knowledge also be." In anticipation, Boyle early committed himself to lifelong celibacy. Invoking Adam's original knowledge, Royal Society secretary Henry Oldenburg proclaimed that the Society's aim was "to raise a Masculine Philosophy," which Society historian Thomas Sprat described as "the Masculine Arts of Knowledge." John Wilkins, leader of the Society founders, maintained that the new knowledge would bring about man's recovery from the Fall, while Robert Hooke wrote a continuation of Bacon's utopia "The New Atlantis" in which the reign of the scientific priesthood on earth was held to correspond with God's governance of the universe: the prophesied reign of the saints.

The rigorously ascetic Isaac Newton devoted a lifetime to the study of Biblical prophecy. He wrote four separate commentaries on Daniel and Revelation and a treatise on "The end of the world, day of judgement, and world to come." Like Bacon, Newton insisted that he was not so much an innovator as a rediscoverer of lost knowledge, that his life's work was a heroic act of recovery. Like Boyle, he speculated about what the kingdom of God would be like. He was convinced that he would be among the "sons of the resurrection" and, according to his first mentor Henry More, "seemed to fancy himself soaring through the heavens . . . filled with a happy throng of saintly companions."

Perhaps the most explicit expression of the masculine Millennium mythology was provided by Joseph Glanvill, another founder and foremost propagandist of the Royal Society, in his treatise "On the Vanity of Dogmatizing." Glanvill began his book with a chapter entitled "What the Man Was," describing the fullness of Adam's original knowledge. "All the faculties of this copy of Divinity," Glanvill exulted, "were as perfect as beauty and harmony in Idea. The senses, the Soul's windows, were without any spot or opacity. . . . Adam needed no spectacles. The acuteness of his natural optics shewed him most of the celestial magnificence and bravery without a Galileo's tube. . . . His naked eyes could reach near as much as of the upper world, as we with all the advantages of the arts. . . . His knowledge was completely built, upon the certain, extempory notice of his comprehensive, unerring faculties. . . . Causes are hid in night and obscuring from us, which were all Sun to him. . . . While man knew no sin, he was ignorant of nothing else."

Alas, Glanvill despaired, because of the Fall "We are not now like the creature we were made, and have not only lost our Maker's image, but our own." But, in keeping with what was now an eight-hundred-year ideological tradition, Glanvill argued that the humble study of nature and the advance of the arts would both partially restore man's original capacities and lay the groundwork for the blessed Millennium itself. He cautioned, however, again following tradition, that "the Woman in us still prosecutes a deceit, like that begun in the Garden." Thus, such efforts promised nothing so long as "our understandings are wedded to an Eve, as fatal as the mother of our Miseries."

From Bacon to Newton, the robust millenarian spirit of the seventeenth century brought the religion of technology from the margins of Western history to centre stage, where it has remained. This spirit was carried forth into the age of enlightenment by other major scientific figures, such as William Whiston, Joseph Priestley, and later Michael Faraday and James Clerk-Maxwell, all religious enthusiasts. More important, this spirit was institutionalized by a cadre of evangelical Newtonians who established a new order of spiritual men, the Freemasons. With roots in Rosicrucianism, the Freemason fraternity inherited a tradition of ritual, symbolism, and lore, a defining interest in the glorification and recovery of ancient knowledge, and a dedication to the advancement and diffusion of the arts. The Freemasons elevated the arts as never before and became the most avid advocates of industrial science.

The opening sentences of the Freemason Constitution resonate with the refrains of redemption. "Adam, our first parent, created after the Image of God, the Great Architect of the Universe, must have had the Liberal Sciences, particularly Geometry, written on his Heart; for even since the Fall we find the principles of it in the Hearts of his offspring." The Fellow-Craft song, sung at the Grand Feast in all lodges, exuberantly proclaims the privileged pursuit of perfection: "Hail Masonry! Thou Craft Divine! / Glory of Earth, from Heaven revealed. / Which dost with Jewels precious sine, / From all but masons' eyes concealed." Like the monks, friars, explorers, magi, and scientists before them, the Freemasons expressly excluded women from their blessed brotherhood, as one Mason put it, "because their presence might insensibly alter the purity of our maxims."

The Freemasons became, in the eighteenth and nineteenth centuries, probably the foremost promoters of what were now called the industrial arts, establishing countless societies throughout Europe and

North America dedicated to that end. But perhaps the most important, and heretofore unexamined, contribution of freemasonry in this regard was its central role in the creation of modern professional engineering and engineering education. From the hands of Freemasons, the mantle of the third age was passed to this new breed of spiritual men, the new Adam, the engineer. The first professional engineering society, the Institution of Civil Engineers in England, was founded by Freemason Thomas Telford, and many of that country's most prominent engineers were practising Freemasons. In France, which pioneered in engineering education and set the standard for engineering professionalism for the world, both the École des Ponts et Chausées and the École Polytechnique were the achievements of Freemasons, in particular Jean Rodolph Perronet and Gaspard Monge. In Prussia, Freemasons likewise played a central role in the formation of the engineering profession, as historian Eric Brose has shown, while in the United States lifelong Freemason Benjamin Franklin was the premier pioneer of technical education and the advancement of the arts.

Thus, via freemasonry, the saintly savants of the Reformation passed the project of redemption through the arts to the engineers, the modern architects of technological transcendence. In time, the engineers elaborated their own secret rituals and exclusively male associations in pursuit of perfection. They were, in the words of two nineteenth-century engineers, the "priests of the new epoch," who were destined to bring about a new day "when every force in nature and every created thing shall be subject to the control of man" — paradise regained. In the nineteenth century this thousand-year-old project was given a new name: "technology." And if the engineers themselves identified their work with destiny, their claims were ratified by a new generation of prophets, from Saint-Simon and Owen to Marx, Bellamy, and Veblen, who placed the engineering enterprise at the core of their transparently millenarian visions. But the true herald of the engineer, and heir to the masculine millenarian tradition, was Auguste Comte, a teacher at the École Polytechnique whose tripartite vision of historical development faithfully reproduced the Joachimite trinitarian scheme.

For Comte too the third age was that of a "transitional period," expressly embodied in the scientific engineer, in which industry and positive science combined with a new "Religion of Humanity" would re-establish the "filiation of man," restoring man to "the normal state," "the definitive form of his existence." Assuming the "function of the prophet," Comte foresaw the "final crisis" followed by the "inevitable" dawning of a "universal order," the "kingdom of the Great Being . . . the advent of which is shown by the whole past to be at hand." Comte wrote of the "direct regenerative efforts made by the priesthood—efforts aiming at the preparation of the normal state and the reconstruction of the West by a worthy glorification of the past."

Despite his theoretical repudiation of theological and metaphysical thinking, Comte reproduced nearly in its entirety the medieval millenarian mythology, explicitly used the medieval church as his model for the future order, aimed at "awakening in all the noble desire of honorable incorporation with the supreme existence," and, for inspiration, daily read from Thomas à Kempis's "Imitation of Christ." Like his monastic forebears, Comte disqualified women from leadership in the new age, assuming that "the natural movement of our industry certainly tends gradually to pass to men the professions long exercised by women"—the recovery of a more perfect order.

⌇ ⌇ ⌇

As Christopher Lasch acutely observed, dangerous psychotic delusions—a celebration of disembodied intelligence and an escapist fantasy of total control and absolute transcendence of the limits imposed on mankind—lie at the core of the modern cult of technology. An alarmingly irrational mentality propels what is seemingly the most rational, and certainly the most consequential, of human projects. As the discussion here demonstrates, this pervasive pathology has its roots in the thousand-year-old religion of technology, the myth of a masculine Millennium.

Over the last century the vocabulary has changed, perhaps, but the dominant mythic themes have not; they have remained intact, and all the more effective because unconscious. In the present age, the explicitly religious vocabulary and consciousness of purpose are rapidly reappearing as men of science and technology have become bolder and more confident of the success of their perfectionist project, the restoration of their divine likeness. Among physicists and biologists the use of the word God and the expression of an explicitly religious purpose have once again become fashionable, if not yet obligatory, while among engineers, the religious quest has moved beyond theory into practice.

In space exploration, they are joining the angels; in artificial reproduction, they are regaining Adam's male-only procreative powers; in artificial intelligence, they are overcoming the curse of toil and the mortal bondage of the body; in genetic engineering, they are becoming once again God's partner in Creation; in cyberspace, they are recovering their rightful dominion over the universe, omniscient and omnipresent. In the Arizona desert, the designers of a totally artificial habitat for use in space have dubbed their project Genesis II, while among Artificial Life designers at the Santa Fe Institute (typically male) programmers routinely describe themselves as gods. "I feel like God," one researcher told Stefan Helmreich, a visiting anthropologist. "In fact, I am God to the universes I create." Predictably, so-called seed programs, which "evolve" new silicon "life" forms, are often called "Adam." In one such program, users can confirm changes in life design by pressing a button that reads "Amen."

But it is not the practitioners alone who are so moved. A thousand years in the making, the religion of technology has become the common faith, shared alike by the designers and by those caught up, undone, or destroyed by their godly designs. The popular expectation of deliverance through technology, whatever the abundantly apparent human and social costs, has become the unspoken orthodoxy, and the shared delusion. Thus, the "advance of the arts" is allowed to proceed apace, without scrutiny, without oversight, without social purpose—without reason. Amply aided and indulged by evangelical corporate promoters (Apple Computer sales personnel are called "evangelists") and self-serving governments, the media, and the military—and supported by an increasingly desperate populace for whom technological transcendence (as escape) appears to be the only possibility—the architects of the new age wield their priest-like authority with Adamic aplomb as they lead us on their accelerating otherworldly adventure. Criticism is dismissed as irrelevant, and irreverent.

Opposition is akin to heresy. Yet for the this-worldly few who remain attuned to more terrestrial trials and tribulations, criticism and opposition have become an urgent imperative. Against this hegemonic system of blind belief, rationality demands resistance—a struggle not for salvation but for survival.

APPENDICES

APPENDIX I

NINETEENTH-CENTURY CONSULTANT TO INDUSTRY SAW AUTOMATION AS WEAPONRY

In the factories for spinning coarse yarn . . . the mule-spinners [skilled workers] have abused their powers beyond endurance, domineering in the most arrogant manner . . . over their masters. High wages, instead of leading to thankfulness of temper and improvement of mind, have, in too many cases, cherished pride and supplied funds for supporting refractory spirits in strikes, wantonly inflicted upon one set of mill-owners after another. . . . During a disastrous turmoil of [this] kind . . . several of the capitalists . . . had recourse to the celebrated machinists . . . of Manchester, requesting them to direct [their] inventive talents . . . to the construction of a self-acting mule. Under assurance of the most liberal encouragement in the adoption of his inventions, Mr. Roberts . . . suspended his professional pursuits as an engineer, and set his fertile services to construct a spinning automation. . . . Thus, the Iron Man, as the operatives fitly call it, sprung out of the hands of our modern Prometheus at the bidding of Minerva—a creation destined to restore order among the industrious classes. . . . This invention confirms the great doctrine already propounded, that when capital enlists science in her service, the refractory hand of labor will always be taught docility.

SOURCES

Andrew Ure, *Philosophy of Manufactures* [London, 1835] (New York: Burt Franklin/Lenox Hill Publishing Corp., 1969), pp.336-68. For more on Andrew Ure (1778-1857), Scottish chemist and first director of Glasgow Observatory, see Maxine Berg, *The Machinery Question and the Making of Political Economy, 1815-1848* (Cambridge, 1980); and *The Dictionary of the History of Ideas*, IV (New York, 1973), p.363.

APPENDIX II

KARL MARX AGAINST THE LUDDITES

The enormous destruction of machinery that occurred in the English manufacturing districts during the first 15 years of this century, chiefly caused by the employment of the power-loom, and known as the *Luddite movement*, gave the anti-jacobin governments of a Sidmouth, a Castlereagh, and the like, a pretext for the most reactionary and forcible measures. It took both time and experience before the workpeople learnt to distinguish between machinery and its employment by capital, and to direct their attacks, not against the material instruments of production, but against the mode in which they are used.

SOURCES

Karl Marx, *Capital*, 3 vols., trans. Samuel Moore, Edward Aveling, and Ernest Untermann (Chicago: Charles H. Kerr & Co., 1906-1909), vol. 1, p.468.

———, *Capital*, vol. 1 in 2 vols., "Introduction" by G.D.H. Cole, trans. Eden and Cedar Paul (New York: Everyman's Library, E.P. Dutton & Co., 1930), vol. 1, p.458.

———, *Capital*, 3 vols., trans. Ben Fowkes (New York: Vintage Books, 1977), vol. 1, pp.554-55.

———, *Capital*, 3 vols., trans. Samuel Moore and Edward Aveling (New York: International Publishers, 1987), vol. 1, p.404.

A Technology Bill of Rights from the International Association of Machinists

APPENDIX III

On April 30, and May 1, 1981, William Winpisinger, then president of the International Association of Machinists and Aerospace Workers (IAM), AFL-CIO, hosted the IAM Scientists and Engineers Conference in New York City. It was chaired by Seymour Melman, Professor of Engineering and Operations Research at Columbia University. The purpose of the event was to bring social scientists from major universities and engineers from large manufacturing corporations into direct dialogue with each other and with top IAM officials and rank and file members, an attendance of about forty people in all.[1] Their assignment was to examine what was happening to the nature of work and employment as the mechanized forms of automation get replaced by those involving computers and robots.

The Technology Bill of Rights was produced as a direct result of the conference.[2] Beyond circulation given the bill directly by the IAM staff, it was also published in the quarterly journal *democracy* (Sheldon Wolin, ed.), New York, Spring 1983, pp.25-27.

ᚵ ᚵ ᚵ

International Association of Machinists Congress hereby amends the National Labor Relations Act, Railway Act, and other appropriate Acts to declare a national labor policy through a New Technology Bill of Rights:

I

New technology shall be used in a way that creates jobs and promotes community-wide and national full employment.

II

Unit labor cost savings and labor productivity gains resulting from the use of new technology shall be shared with workers at the local enterprise level and shall not be permitted to accrue excessively or exclusively for the gain of capital, management, and shareholders. Reduced work hours and increased leisure time made possible by new technology shall result in no loss of real income or decline in living standards for workers affected at the local level.

III

Local communities, the states, and the nation have a right to require employers to pay a replacement tax on all machinery, equipment, robots, and production systems that displace workers and cause unemployment, thereby decreasing local, state, and federal revenues.

IV

New technology shall improve the conditions of work and shall enhance and expand the opportunities for knowledge, skills and compensation of workers. Displaced workers shall be entitled to training, retraining, and subsequent job placement or re-employment.

V

New technology shall be used to develop and strengthen the U.S. industrial base, consistent with full employment goals and national security requirements, before it is licensed or otherwise exported abroad.

VI

New technology shall be evaluated in terms of worker safety and health and shall not be destructive of the workplace environment, nor shall it be used at the expense of the community's natural environment.

VII

Workers, through their trade unions and bargaining units, shall have an absolute right to participate in all phases of management deliberations and decisions that lead or could lead to the introduction of new technology or the changing of the workplace system design work process, and procedures for doing work, including the shutdown or transfer of work, capital, plants, and equipment.

VIII

Workers shall have the right to monitor control room centers and control stations, and the new technology shall not be used to monitor, measure or otherwise control the work practices and work standards of individual workers at the point of work.

IX

Storage of an individual worker's personal data and information file by the employer shall be tightly controlled, and the collection and/or release and dissemination of information with respect to race, religion, or political activities and beliefs, records of physical and mental health disorders and treatment, records of arrests and felony charges or convictions, information concentrating intentional and private family matters, and information regarding an individual's financial condition or credit worthiness, shall not be permitted, except in rare circumstances related to health, and then only after consultation with a family or union-appointed physician, psychiatrist, or member of the clergy. The right of an individual worker to inspect his or her personal file shall at all times be absolute and open.

X

When the new technology is employed in the production of military goods and services, workers, through their trade unions and bargaining agents, have a right to bargain with management over the establishment of Alternative Production Committees, which shall design ways to adapt that technology to socially useful production in the civilian sector of the economy.

NOTES

1 David Noble was a participant.
2 "I drafted this version of 'The Technology Bill of Rights,' with considerable input from Seymour Melman and the other attendees of the 'Scientists and Engineers Conference.'" From Harley Shaiken, *Work Transformed: Automation and Labor in the Computer Age* (New York: Holt, Rinehart, and Winston, 1984), notes to Chapter 8, p.296.

Appendix IV

"Starvin' in Paradise" with the New Technology

by Dick Greenwood,
Special Assistant to the International President,
International Association of Machinists and Aerospace Workers

The relationship of worker and machine is being radically transformed. Heretofore, machines, by and large, have replaced muscle in the world of work. Human labor, however, has remained the principal factor of production in the creation of wealth, albeit, it has never been adequately recognized or rewarded as such.

But there are signs the new technology currently coming down will relegate labor, as a *factor of production*, to secondary status. When robots build robots that cut, shape, weld, paint, assemble, and load and unload autos and appliances, then, brothers and sisters, we're talking about the primacy of capital, because the labor input into that kind of production has been drastically reduced or eliminated from the shop floors. Academics begin talking about the productivity of *capital*, not labor.

The new technology not only replaces human muscle, it is replacing human brainpower and the human nervous system. Artificial intelligence has long plagued management, but it is no laughing matter to learn that *real* artificial intelligence is being produced in university and corporate laboratory machines. Don't worry about the machines' IQs, we all know management has seldom given us credit for having an IQ, anyway, and, after Frederick Taylor, never wanted us to have much intelligence—just enough to receive and respond to commands from on high, like "speed it up," "don't ask questions," "quit quibbling about pay," "you're fired," and maybe, "you're hired."

Now, GE and its Business Roundtable collaborators have not only installed Ronnie the Robot in the White House (it's not hard to guess *his* IQ), they're stalking us in the workplaces, too.

Forget about that nice, clean, pink and white collar work we're supposed to get when technology drives us from the production floor, skilled machine shops, and warehouse, because the New Technology is already emptying people out of whole office buildings. Apply for a job there, and you'll get a short spin through the revolving door by an exodus already under way. Word processors with memory systems and automatic printers make up the new wave there. One operator can handle a dozen of them if properly "Taylorized." Similarly, new computerized machining centers and work cells for loading, cutting, grinding, milling, polishing, shaping, transfer, and assembly operations on the manufacturing floor eliminate crafts, skills, and people, too.

Even if we escape "technological unemployment" for a few years, we're going to find our workplaces organized differently, the content of our jobs changed and our skills reduced, new and obsolete job classifications, new remote management control systems and control centers. Grievances are going to mount, contract language is going to be vague or non-existent to deal with the effects, and the jargon and language management starts laying on us will be an absolute snow job — an obfuscator's dream.

This whole change is just beginning. It's going to accelerate, fuelled primarily by American management's panic over the mess it has made of things in the current depression. Management sees the new technology as a quick fix for its greedy, short-sighted decisions and bungling investment policies. It also sees it as a technological end run around trade unions and collective bargaining procedures. Of course, our employers are getting the money to pay for all this new stuff through those generous corporate tax cuts and tax credits handed to them last year by the Reaganites and the Congress.

In view of all this, the IAM called together a group of shop floor stewards, a few Local and District Officers, and some savvy production members who don't hold office, along with some corporate engineers and a similar number of academic industrial engineers and scientists, locked them up together for a week and cussed and discussed the implications of what's happening.

After everyone was gone, the IAM members drew up a Technological Bill of Rights to present to the public and create an awareness of the social effects of technology, as well as its economic and technical aspects, which get all the notice now. We also drew up a list of specific recommendations, to present to the union as a whole, for more discussion and consideration. Each document is available on request.

At this point, the objective is not to block the new technology, but to control its rate and manner of introduction, in order that it is adapted to labor's needs and serves people, rather than our being servile to it or its victims. It can go either way, and it's headed the wrong way right now.

Let it be known that if ever we needed a shorter work-week, without loss of real pay, now is the time to go after it. If we don't, we're going to have leisure time we never dreamed of but, just like those millions of us out of work now, that leisure time ain't going to be compensated.

As one critic has already put it, we'll be starving in Paradise.

IAM SCIENTISTS AND ENGINEERS CONFERENCE RECOMMENDATIONS:

1. The Norwegian Metal Workers Technology Control program be used as a model for the IAM Legislative and Local Union program.
2. Each District and Local Lodge establish a Technology Control Committee and elect or appoint Technology Control Stewards. Technology Control Committee to include members of Safety and Health Committee and at least two Bargaining Committee members.
3. Local and District Lodge Technology Control Committees establish liaison with company professional engineering staff, in an effort to gain their confidence and take advantage of their specialized knowledge and information, while developing a technology control program.
4. Local and District Technology Control Committee work closely with Bargaining Committee to give very high priority to Technology Control as a bargaining issue and that contract language be developed to protect IAM members through advance notice and mandatory consultation, prior to company decision to implement or install new technology, equipment, or systems. Realizing that whenever New Technology is introduced somewhere, somebody is going to be displaced or go out the door, the Technology Committee should assess and evaluate the introduction of New Technology on a company-wide or system-wide basis, not just on a departmental or plant-wide basis.
5. When management gives advance notice and consults the Local or District Lodge concerning introduction of new technology, the Local or District Lodge should attempt to persuade the Company

    a) that members of the Technology Control Committee accompany management as it shops for software, so that the union can have thorough knowledge of and a voice in the design of the systems and include the shop floor workers "in the loop."
    b) that *before* new technology hardware and equipment is purchased, leased, or installed, that management seriously consider the alternative of manufacturing such equipment. In the case of subcontracting out work, every effort should be made to keep the work in the plant or in the company with a bargaining unit.
    c) that all new technology proposals advanced by management be presented to the union in language easily understood by all persons without special knowledge of the technology concerned.
    d) that all programming, editing, operator, and servicing jobs required by introduction of new technology be included in the bargaining unit.
    e) that the union has a right to monitor control room and control center operations.

- f) that retraining and job classifications for workers displaced or transferred by new technology be agreed upon before company is locked into purchase, lease, or contract for new technology and that training and retraining will be paid for by the company.
- g) that any or all agreements made with the company concerning new technology be properly negotiated and written into the contract or be clearly stated in interim letters of agreement.
- h) that disputes over the introduction or use and misuse of New Technology is a strikable issue.

6. The Grand Lodge design and conduct a pilot school to train Technology Control Stewards and Technology Control Committee members in both software and hardware associated with the New Technology.
7. The Grand Lodge make every effort to influence the AFL-CIO, through its various departments, to adopt and implement a New Technology Program along the lines proposed by the Conference.
8. The Grand Lodge impanel a group of Scientists and Engineers, who are intimately involved with and have expert knowledge of the New Technology and convene them for the purpose of evaluating the new technology, not only from technical and economic considerations, but with an emphasis on safety, health, and social consequences, and further, that the Grand Lodge initiate this project by first exploring it with the Union of Concerned Scientists.
9. The Grand Lodge embody its national legislative program in a Technology Bill of Rights for American workers and this program be included in the IAM's Rebuilding America program.

Appendix V

Lord Byron Speaks Against a Bill to Introduce the Death Penalty for Machine-Breaking

Speech in the House of Lords, February 27, 1812

Lord Byron: My Lords, The subject now submitted to your Lordships for the first time is by no means new to the country. I believe it had occupied the serious thoughts of all descriptions of persons long before its introduction to the notice of the legislature, whose interference alone could be of real service. As a person in some degree connected with the suffering county, though a stranger not only to this House in general but to almost every individual whose attention I presume to solicit, I must claim some portion of your Lordships' indulgence, whilst I offer a few observations on a question in which I confess myself deeply interested.

To enter into any detail of the riots would be superfluous; the House is already aware that every outrage short of actual bloodshed has been perpetrated, and that the proprietors of the frames obnoxious to the rioters, and all persons supposed to be connected with them, have been liable to insult and violence. During the short time I recently passed in Nottinghamshire, not twelve hours elapsed without some fresh act of violence; and on the day I left the county I was informed that forty frames had been broken the previous evening, as usual, without resistance and without detection.

Such was then the state of that county, and such I have reason to believe it to be at this moment. But whilst these outrages must be admitted to exist to an alarming extent, it cannot be denied that they have arisen from circumstances of the most unparalleled distress; the perseverance of these miserable men in their proceedings tends to prove that nothing but absolute want could have driven a large, and once honest and industrious, body of the people into the commission of excesses so hazardous to themselves, their families, and the community. At the time to which I allude, the town and county were burdened with large detachments of the military; the police was in motion, the magistrates assembled; yet all the movements, civil and military, had led to — nothing. Not a single instance had occurred of the apprehension of any real delinquent actually taken in the fact, against whom there existed legal evidence sufficient for conviction. But the police, however useless, were by no means idle; several notorious delinquents had been detected — men, liable to conviction, on the clearest evidence, of the capital crime of poverty; men, who had been nefariously guilty of lawfully begetting several children, whom, thanks to the times! they were unable to maintain. Considerable injury has been done to the proprietors of the improved frames. These machines were to them the advantage, inasmuch as they superseded the necessity of employing a number of workmen, who were left in consequence to starve. By the adoption of one species of frame in particular, one man performed the work of many, and the superfluous labourers were thrown out of employment. Yet it is to be observed, that the work thus executed was inferior in quality, not marketable at home, and merely hurried over with a view to exportation. It was called, in the cant of the trade, by the name of "Spider-work." The rejected workmen, in the blindness of their ignorance, instead of rejoicing at these improvements in arts so beneficial to mankind, conceived themselves to be sacrificed to improvements in mechanism. In the foolishness of their hearts they imagined that the maintenance and well-doing of the industrious poor were objects of greater consequence than the enrichment of a few individuals by any improvement, in the implements of the trade, which threw the workmen out of employment and rendered the labourer unworthy of his hire. And it must be confessed that although the adoption of the enlarged machinery in that state of our commerce which the country once boasted might have been beneficial to the master without being detrimental to the servant; yet, in the present situation of our manufactures, rotting in warehouses, without a prospect of exportation, with the demand for work and workmen equally diminished, frames of this description tend materially to aggravate the distress and discontent of the disappointed sufferers. But the real cause of these distresses and consequent disturbances lies deeper. When we are told that these men are leagued together not only for the destruction of their own comfort, but of their very means of subsistence, can we forget that it is the bitter policy, the destructive warfare of the last eighteen years, which had destroyed their comfort, your comfort, all men's comfort? That policy, which originating with "great statesmen now no more," has survived the dead to become a curse on the living, unto the third and fourth generation! These men never destroyed their looms till they were become useless, worse than useless; till they were become actual impediments to their exertions in obtaining their daily bread. Can you, then, wonder that in times like these, when bankruptcy, convicted fraud, and imputed felony are found in a station not far beneath that of your Lordships, the lowest though once most useful portion of the people, should forget their duty in their distresses, and become only less guilty than one of their representatives? But while the exalted offender can find means to baffle the law, new capital punishments must be devised, new snares of death must be spread for the wretched mechanic, who is famished into guilt. These men were willing to dig, but the spade was in other hands; they were not ashamed to beg, but there was none to relieve them; their own means of subsistence were cut off, all other employments preoccupied; and their excesses, however to be deplored and condemned, can hardly be subject of surprise.

It has been stated that the persons in the temporary possession of frames connive at their destruction; if this be proved upon inquiry, it were necessary that such material accessories to the crime should be principals in the punishment. But I did hope, that any measure proposed by His Majesty's Government for your Lordships' decision would have had conciliation for its basis; or, if that were hopeless, that some previous inquiry, some deliberation, would have been deemed requisite, not that we should have been called at once, without examination and without cause, to pass sentences by wholesale, and sign death-warrants blindfold. But, admitting that those men had no cause of complaint; that the grievances of them and their employers were alike groundless; that they deserved the worst; what inefficiency, what imbecility has been evinced in the method chosen to reduce them! Why were the military called out to be made a mockery of, if they were to be called out at all? As far as the difference of seasons would permit, they have merely parodied the summer campaign of Major Sturgeon; and, indeed, the whole proceedings, civil and military, seemed on the model of those of the mayor and corporation of Garrett. Such marchings and countermarchings!—from Nottingham to Bullwell from Bullwell to Banford, from Banford to Mansfield! And when at length the detachments arrived at their destination, in all "the pride, pomp, and circumstance of glorious war," they came just in time to witness the mischief which had been done, and ascertain the escape of the perpetrators, to collect the *spolia opima* in the fragments of broken frames, and return to their quarters amidst the derision of old women, and the hootings of children. Now though in a free country it were to be wished that our military should never be too formidable, at least to ourselves, I cannot see the policy of placing them in situations where they can only be made ridiculous. As the sword is the worst argument that can be used, so should it be the last. In this instance it has been the first; but providentially as yet only in the scabbard. The present will, indeed, pluck it from the sheath; yet had proper meetings been held in the earlier stages of these riots, had the grievances of these men and their masters (for they also had their grievances) been fairly weighed and justly examined, I do think that means might have been devised to restore these workmen to their avocations, and tranquillity to the county. At present the county suffers from the double affliction of an idle military and a starving population. In what state of apathy have we been plunged so long, that now for the first time the House has been officially apprised of these disturbances? All this has been transacting within 130 miles of London; and yet we, "good easy men, have deemed full sure our greatness was a ripening," and have sat down to enjoy our foreign triumphs in the midst of domestic calamity. But all the cities you have taken, all the armies which have retreated before your leaders, are but paltry subjects of self-congratulation, if your land divides against itself, and your dragoons and your executives must be let loose against your fellow-citizens. You call these men a mob, desperate, dangerous, and ignorant; and seem to think that the only way to quiet the *Bellua multorum capitum* is to lop off a few of its superfluous heads. But even a mob may be better reduced to reason by a mixture of conciliation and firmness, than by additional irritation and redoubled penalties. Are we aware of our obligation to a mob? It is the mob that labour in your fields, and serve in your houses, that man your navy and recruit your army, that have enabled you when neglect and calamity have driven them to despair! You may call the people a mob; but do not forget that a mob too often speaks the sentiments of the people. And here I must remark, with what alacrity you are accustomed to fly to the succour of your distressed allies, leaving the distressed of your own country to the care of Providence — or — the parish. When the Portuguese suffered under the retreat of the French, every arm was stretched out, every hand was opened, from the rich man's largess to the widow's mite, all was bestowed, to enable them to rebuild their villages and replenish their granaries. At this moment when thousands of misguided but most unfortunate fellow-countrymen are struggling with the extremes of hardships and hunger, as your charity began abroad it should end at home. A much less sum, a tithe of the bounty bestowed on Portugal, even if those men (which I cannot admit without inquiry) could not have been restored to their employments, would have rendered unnecessary the tender mercies of the bayonet and the gibbet. But doubtless our friends have too many foreign claims to admit a prospect of domestic relief, though never did such objects demand it. I have traversed the seat of war in the Peninsula, I have been in some of the most oppressed provinces of Turkey, but never under the most squalid wretchedness I have seen since my return in the very heart of Christian country. And what are your remedies? After months of inaction, and months of action worse than inactivity, at length comes forth the grand specific, the never-failing nostrum of all state physicians, from the days of Draco to the present time. After feeling the pulse and shaking the head over the patient, prescribing the usual course of warm water and bleeding — the warm water of your mawkish police, and the lancets of your military — these convulsions must terminate in death the sure consummation of the prescriptions of all political Sangrados. Setting aside the palpable injustice and the certain inefficiency of the Bill, are there not capital punishments sufficient in your statutes? Is there not blood enough upon your penal code, that more must be poured forth to ascend to heaven and testify against you? How will you carry the Bill into effect? Can you commit a whole county to their prisons? Will you erect a gibbet in every field and hang up men like scarecrows? or will you proceed (as you must to bring this measure into effect) by decimation? place the county under martial law? depopulate and lay waste all around you? and restore Sherwood Forest as an acceptable gift to the crown, in its former condition of a royal chase and an asylum for outlaws? Are these the remedies for a starving and desperate populace? Will the famished wretch who has braved your bayonets be appalled by your gibbets? When death is a relief, and the only relief it appears that you will afford him, will he be dragooned into tranquillity? Will that which could not be affected by your grenadiers be accomplished by your executioners? If you proceed by the forms of law, where is your evidence? Those who have refused to impeach their accomplices when transportation only was the punishment, will hardly be tempted to witness against them when death is the penalty. With all due deference to the noble lords opposite, I think a little investigation, some previous inquiry, would induce even them to change their purpose. That most favourite state measure, so marvellously efficacious in many and recent instances, temporising, would not be without its advantages in this. When a proposal is made to emancipate or relieve, you hesitate, you deliberate for years, you temporise and tamper with the minds of men; but a death-bill must be passed offhand, without a thought of the consequences. Sure I am, from what I have heard, and from what I have seen, that to pass the Bill under all the existing circumstances, without inquiry, without deliberation, would only be to add injustice to irritation, and barbarity to neglect. The framers of such a Bill must be content to inherit the honours of that Athenian law-giver whose edicts were said to be written not in ink but in blood. But suppose it passed; suppose one of these men as I have seen them—meagre with famine, sullen with despair, careless of a life which you Lordships are perhaps about to value at something less than the price of a stocking frame;—suppose this man surrounded by the children for whom he is unable to procure bread at the hazard of his existence, about to be torn for ever from a family which he lately supported in peaceful industry, and which it is not his fault that he can no longer so support;—suppose this man, and there are ten thousand such from whom you may select your victims, dragged into court, to be tried for this new offence, by this new law, still, there are two things wanting to convict and condemn him, and these are, in my opinion, twelve butchers for a jury, and a Jefferies for a judge!

(The bill passed, with three votes against—ed.)

Letter from Lord Byron to Lord Holland
8, St James's Street, February 25, 1812

My Lord. With my best thanks, I have the honour to return the Notts. letter to your Lordship. I have read it with attention, but do not think I shall venture to avail myself of its contents, as my view of the question differs in some measure from Mr. Coldham's. I hope I do not wrong him, but *his* objections to the bill appear to me to be founded on certain apprehensions that he and his coadjutors might be mistaken for the *"original advisers"* (to quote him) of the measure. For my own part I consider the manufacturers as a much injured body of men, sacrificed to the views of certain individuals who have enriched themselves by those practices which have deprived the frame-workers of employment. For instance; — by the adoption of a certain kind of frame, one man performs the work of seven — six are thus thrown out of business. But it is to be observed that the work thus done is far inferior in quality, hardly marketable at home, and hurried over with a view to exportation. Surely, my Lord, however we may rejoice in any improvement in the arts which may be beneficial to mankind, we must not allow mankind to be sacrificed to improvements in mechanism. The maintenance and well-doing of the industrious poor is an object of greater consequence to the community than the enrichment of a few monopolists by any improvement in the implements of trade, which deprives the workman of his bread, and renders the labourer "unworthy of his hire."

My own motive for opposing the bill is founded on its palpable injustice, and its certain inefficacy. I have seen the state of these miserable men, and it is a disgrace to a civilized country. Their excesses may be condemned, but cannot be subject of wonder. The effect of the present bill would be to drive them into actual rebellion. The few words I shall venture to offer on Thursday will be founded upon these opinions formed from my own observations on the spot. By previous inquiry, I am convinced these men would have been restored to employment, and the county to tranquility. It is, perhaps, not yet too late, and is surely worth the trial. It can never be too late to employ force in such circumstances. I believe your Lordship does not coincide with me entirely on this subject and most cheerfully and sincerely shall I submit to your superior judgment and experience, and take some other line of argument against the bill or be silent altogether, should you deem it more advisable. Condemning, as every one must condemn, the conduct of these wretches, I believe in the existence of grievances which call rather for pity than punishment. I have the honour to be, with great respect, my Lord, your Lordship's

Most obedient and obliged servant,
Byron

P.S. — I am a little apprehensive that your Lordship will think me too lenient towards these men, and half a *frame-breaker myself*.

Sources

The speech appears as an appendix in Ernst Toller, *The Machine Wreckers: A Drama of the English Luddites in a Prologue and Five Acts* (London, 1923), pp.105-13. The letter from Lord Byron (George Gordon Byron, 6th Lord of Byron, 1788-1824) is from his *Selected Poems and Letters* (New York: New York University Press, 1977), pp.448-49.

Appendix VI

An Exchange between Norbert Wiener, Father of Cybernetics, and Walter Reuther, UAW President

Wiener's Letter to Reuther

South Tamworth, August 13, 1949

Walter Reuther
Union of Automobile Workers
Detroit, Michigan

Dear Mr. Reuther:

First, I should like to explain who I am. I am Professor of Mathematics at the Massachusetts Institute of Technology, and am the author of the recently published book, *Cybernetics* (Wiley and Sons and the Technology Press). As you will see, if you know of this book, I have been interested for a long time in the problem of automatic machinery and its social consequences. These consequences seem to me so great that I have made repeated attempts to get in touch with the Labor Union movement, and to try to acquaint them with what may be expected of automatic machinery in the near future. This situation has been brought to a head by the fact that I have been approached recently by one of the leading industrial corporations with the view to advising them as to whether to go into the problem of making servo-mechanisms, that is, artificial control mechanisms, as part of their extended program.

Technically I have no doubt what direction my advice should take. My technical advice would be to construct an inexpensive small scale, high speed computing machine, together with adequate apparatus for putting the readings of photo-electric cells, thermometers, and other instruments into the machine as numerical data, and for putting numerical out-put data into the motion of shafts and other out-put apparatus. The position of these output shafts should be monitored by proper sense organs, and be put back into the machine as part of the information on which it is to work.

The detailed development of the machine for particular industrial purpose is a very skilled task, but not a mechanical task. It is done by what is called "taping" the machine in the proper way, much as present computing machines are taped. This apparatus is extremely flexible, and susceptible to mass production, and will undoubtedly lead to the factory without employees; as for example, the automatic automobile assembly line. In the hands of the present industrial set-up, the unemployment produced by such plants can only be disastrous. I would give a guess that a critical situation is bound to arise under any condition in some ten to twenty years; but that if war should make the replacement of labor mobilized into the services an immediate necessity, we should probably have a concentrated effort put into this work which might well lead to large scale industrial unemployment within two years.

I do not wish personally to be responsible for any such state of affairs. I have, therefore, turned down unconditionally the request of the industrial company which has tried to consult me. However, it is manifestly not enough to take a negative attitude on this. If I do not put this information in the hands of the industrialists, it is merely a question of time when so obvious a method of procedure will be urged upon them by other people.

Therefore, the procedure which I shall follow depends finally upon whether I can get you and the labor interests you represent to pay serious attention to this serious situation. I have tried to do this in the past without success; and I do not blame you people for it, but since then there has been a turn-over in personnel among you and the present group of labor leaders seem to have transcended the point of view of the shop to a sufficient extent to make it worthwhile for me to make an appeal to you again.

What I am proposing is this. First, that you show a sufficient interest in the very pressing menace of the large-scale replacement of labor by machine on the level not of energy, but of judgment, to be willing to formulate a policy towards this problem. In particular, I do not think it to be at all foolish for you to steal a march upon the existing industrial corporations in this matter; and while taking a part in production of such machines to secure the profits in them to an organization dedicated to the benefit of labor. It may be on the other hand, that you think the complete suppression *[sic]* of these ideas is in order. In either case, I am willing to back you loyally, and without any demand or request for personal returns in what I consider will be a matter of public policy. I wish to warn you, however, that my own passiveness in this matter will not, on the face of it, produce a passiveness in other people who may come by the same ideas, and that these ideas are very much in the air.

If you determine that the matter does not deserve your serious consideration, you will leave me in a very difficult position. I do not wish to contribute in any way to selling labor down the river, and I am quite aware that any labor, which is in competition with slave labor, whether the slaves are human or mechanical, must accept the conditions of work of slave labor. For me merely to remain aloof is to make sure that the development of these ideas will go into other hands which will probably be much less friendly to organized labor.

Under these circumstances, I should probably have to try to find some industrial group with as liberal and honest a labor policy as possible and put my ideas in their hands. I must confess, however, that I know of no group which has at the same time a sufficient honesty of purpose to be entrusted with these developments, and a sufficiently firm economic and social position to be able to hold these results substantially in their own hands.

I have a book (*The Human Use of Human Beings*) which will be forthcoming with Houghton-Mifflin next spring which will bring these ideas to a head. If you so wish, I shall send you copies of the relevant chapters.

Naturally, I do not expect you to take these matters on my momentary say-so. If you show sufficient interest to be willing to push the matter further, I shall be glad to put my ideas both technical and social at your disposal, so that you will be able to judge them better.

Sincerely yours,
Norbert Wiener
Department of Mathematics
Massachusetts Institute of Technology
Cambridge 39, Massachusetts

Reuther's Response to Wiener

BA050 DEC275
  DE.LLR283 PD=WUX DETROIT SCH 17 317P=
  PROFESSOR NORBERT WIENER= =SOUTH TAMWORTH NHAMP=
  DEEPLY INTERESTED IN YOUR LETTER. WOULD LIKE TO DISCUSS IT WITH YOU AT EARLIEST OPPORTUNITY FOLLOWING CONCLUSION OF OUR CURRENT NEGOTIATIONS WITH FORD MOTOR COMPANY. WILL YOU BE ABLE TO COME TO DETROIT=
  =WALTER P REUTHER PRESIDENT UAW CIO
  (352 PM AUG 17 49)

A Note on the Author

Historian David F. Noble has been writing about the social development of science and technology for two decades. He is the author of *America by Design: Science, Technology and the Rise of Corporate Capitalism* (1977); *Forces of Production: A Social History of Industrial Automation* (1984); and *A World Without Women: The Christian Clerical Culture of Western Science* (1992).

Noble's work has examined how science and technology develop as products not only of accumulated knowledge and skills but also of social power and conflict. He has also stepped out of the conventional academic role to become a social critic and activist, working with rank and file groups in several industries in the struggle over new technology. He is the co-founder, with Ralph Nader, of the National Coalition for Universities in the Public Interest.

Noble taught for nine years at the Massachusetts Institute of Technology. After he was unjustly fired in 1984 for his ideas and his actions in support of those ideas, he successfully brought a suit against MIT to obtain and make public the documentary record of his political firing; on the basis of this record the American Historical Association subsequently condemned MIT for the firing.

After his experience at MIT Noble worked with the Smithsonian Institution in Washington as Curator of Industrial Automation and Labor. He put together plans for an exhibit called "Automation Madness," which included Enoch's hammer, the only Luddite sledgehammer still in existence. This exhibit proved too hot for the Smithsonian to handle. They fired David Noble and sent the hammer back to England.

Noble is now professor of history at York University, Toronto. He is currently working on an exploration of the religious roots of the masculine culture of science and technology entitled *The Religion of Technology: The Myths of a Masculine Millennium*. He recently co-founded the Canadian Forum on Higher Education in the Public Interest.

OTHER TITLES OF RELATED INTEREST FROM BETWEEN THE LINES

Shifting Time:
Social Policy and the Future of Work
Armine Yalnizyan, T. Ran Ide, and Arthur J. Cordell
Introduction by Jamie Swift
Afterword by Ursula M. Franklin
$12.95 paper, 0-921284-91-8

Wheel of Fortune:
Work and Life in the Age of Falling Expectations
Jamie Swift
$18.95 paper, 0-921284-89-6

Capitalism Comes to the Backcountry:
The Goodyear Invasion of Napanee
Bryan D. Palmer
$15.95 paper, 0-921284-87-X